MS-DOS® SmartStart

LoriLee M. Sadler
Indiana University, Bloomington

Bill Weil

COLLEGE

Publisher: David P. Ewing

Associate Publisher: Rick Ranucci

Product Development Manager: Thomas A. Bennett

Operations Manager: Sheila Cunningham

Book Designer: Scott Cook

Production Team: Claudia Bell, Paula Carroll, Brad Chinn, Jodie Cantwell, Laurie Casey, Carla Hall-Batton, Caroline Roop, Linda Seifert, Johnna VanHoose

About the Author

LoriLee Sadler teaches introductory computer courses, using both Macintosh and IBM computers, to more than 1,700 students each year at Indiana University. Since 1989, she has designed and developed curriculum for noncomputer science majors and worked with other computer science faculty to create a computing program for nonmajors; this program is being used as a model at many large universities. In January 1993, Sadler took a senior administrative position in the Associate Vice President's office for Information Technology as the pedagogical expert. She continues to teach in the department of computer science and is currently finishing her third Macintosh applications text with coauthor Alan L. Eliason. Sadler holds a B.A. from Northeast Missouri State University and an M.A. from Indiana University.

Acknowledgments

I'd like to thank Carol Crowell, title manager of Que College of Prentice Hall Computer Publishing, for envisioning this project and working to bring this series to fruition. Jeannine Freudenberger, senior editor, whose home and office phone numbers I now have memorized, has been invaluable in working out the logistics of this project and ensuring consistency and quality in the writing. Copy editors Sara Allaei, Fran Blauw, and Jill Bond, along with Betsy Brown, editorial assistant, worked furiously to copy edit this text and played the role of devil's advocate like fine stage performers. This has been a fun project and I can attribute that fact to everyone who has been involved. Thanks!

Title Manager
Carol Crowell

Senior Editor
Jeannine Freudenberger

Editors
Barb Colter
Sara Allaei
Fran Blauw
Jill Bond
Virginia Noble

Editorial Assistant
Elizabeth D. Brown

Formatter
Jill Stanley

Trademark Acknowledgments

Composed in ITC Garamond and MCPdigital by Que Corporation

Give Your Computer Students a SmartStart on the Latest Computer Applications

Que's SmartStart series from Prentice Hall Computer Publishing combines the experience of the Number 1 computer book publisher in the industry with the pedagogy you've come to expect in a textbook.

SmartStarts cover just the basics in a format filled with plenty of step-by-step instructions and screen shots.

Each SmartStart chapter ends with a "Testing Your Knowledge" section that includes true/false, multiple choice, and fill-in-the-blank questions; two or three short projects; and two long projects. The long projects are continued throughout the book to help students build on skills learned in preceding chapters.

Each SmartStart comes with an instructor's manual featuring additional test questions, troubleshooting tips, and additional exercises. This manual will be available both on disk and bound.

Look for the following additional SmartStarts:

Word for Windows SmartStart	1-56529-204-9
Excel 4 for Windows SmartStart	1-56529-202-2
Windows 3.1 SmartStart	1-56529-203-0
WordPerfect 5.1 SmartStart	1-56529-246-4
Lotus 1-2-3 SmartStart (covers 2.4 and below)	1-56529-245-6
dBASE IV SmartStart	1-56529-251-0

For more information call:

1-800-428-5331

or contact your local Prentice Hall College Representative

Contents at a Glance

Introduction 1

1 Understanding Computer Technology 5

2 The Anatomy of DOS 29

3 A First Look at DOS Commands 57

4 Understanding and Using Directories 93

5 Maintaining Files 113

6 Beyond the Ten Commandments.................... 147

7 Customizing DOS 173

A Setup and Installation 197

B Errors Great and Small 211

Index 237

Table of Contents

Introduction 1

    What Is MS-DOS? 1
    What Does This Book Contain? 2
    What Hardware Do You Need To Run MS-DOS? 3
    Conventions Used in This Book 3

1   Understanding Computer Technology 5

    Objectives 6
    A Bit of History… 8
    Objective 1: To Differentiate between Hardware and Software 8
    Objective 2: To Understand the Different Types of Displays 10
    Objective 3: To Use the Extended Keyboard Layout 12
    Objective 4: To Understand the Relationship
        among Computer Components 15
    Objective 5: To Understand the Basics of Data Processing 22
    Chapter Summary 24
    Testing Your Knowledge 24

2   The Anatomy of DOS 29

    Objectives 29
    Objective 1: To Understand the Functions
        of an Operating System 31
    Objective 2: To Learn the Conventions for Naming DOS Files 34
    Objective 3: To Understand How DOS
        Manages Files and Applications 35
    Objective 4: To Perform a Cold and Warm Boot 38
    Objective 5: To Access and Use the DOS Shell 40
    Chapter Summary 54
    Testing Your Knowledge 54

3 **A First Look at DOS Commands** 57

Objectives 57

Objective 1: To Understand the Parts of a DOS Command 59

Objective 2: To Issue Commands from the DOS Prompt 68

Objective 3: To Use the DIR Command 70

Objective 4: To Use Wild Cards Appropriately 73

Objective 5: To Format High- and Low-Density Disks 75

Objective 6: To Use the COPY Command 85

Objective 7: To Use Other Basic DOS Commands 86

Chapter Summary 88

Testing Your Knowledge 89

4 **Understanding and Using Directories** 93

Objectives 93

Objective 1: To Understand the Concept of Directories 94

Objective 2: To Navigate within the Directory Structures 95

Objective 3: To Understand DOS Paths 98

Objective 4: To Understand and Explore Subdirectories 102

Objective 5: To Manage Directories on Your Hard Drive 105

Chapter Summary 108

Testing Your Knowledge 109

5 **Maintaining Files** 113

Objectives 113

Objective 1: To Label Disks and Files 115

Objective 2: To Use Absolute and Relative Paths 115

Objective 3: To View Files in Other Directories 116

Objective 4: To Copy Files 117

Objective 5: To Delete Files 119

Objective 6: To Use DISKCOPY To Copy Disks 120

Objective 7: To Avoid Data Loss 122

Objective 8: To Recover Deleted Files 126

Chapter Summary 142

Testing Your Knowledge 143

6 **Beyond the Ten Commandments** 147

Objectives 147

Objective 1: To Determine the Amount
     of Memory Installed in Your Computer 149

Objective 2: To Examine the Content
and Space Availability on a Disk 149
Objective 3: To Clear the Screen 154
Objective 4: To Determine the Version of DOS
Installed on Your Computer 155
Objective 5: To Turn the VERIFY Option On and Off 156
Objective 6: To Determine the Volume Label of a Disk 157
Objective 7: To Change the Volume Label of a Disk 158
Objective 8: To Understand the DOS RECOVER Command 159
Objective 9: To Learn How To Redirect
Output to Various Devices 161
Objective 10: To Use Pipes and Filters
To Customize DOS Commands 165
Chapter Summary 168
Testing Your Knowledge 168

7   Customizing DOS 173
Objectives 173
Objective 1: To Use the PATH Command 174
Objective 2: To Change the DOS Prompt
by Using the PROMPT Command 177
Objective 3: To Understand Batch Files 179
Objective 4: To Create a Batch File 181
Objective 5: To Understand the AUTOEXEC.BAT File 184
Objective 6: To Create an AUTOEXEC.BAT File 188
Objective 7: To Understand the CONFIG.SYS File 189
Chapter Summary 192
Testing Your Knowledge 192

A   Setup and Installation 197

B   Errors Great and Small 211
What Are Error Messages? 211
How Serious Are Error Messages? 212
Interpreting Error Messages 212

Index 237

Introduction

*M*S-DOS *SmartStart* describes the connection between personal computer hardware and the disk operating system and explains to beginning users all the most frequently used DOS commands from the command line and the DOS Shell. After you become familiar with DOS's commands and features, you can use the figures presented in this book for quick reference.

What Is MS-DOS?

By itself, a computer is just a box and a screen. Ultimately, you are faced with the task of bringing the computer to life. This task requires that you master the computer's operating system, which for most of us is MS-DOS.

MS-DOS (or to most PC users, just DOS) is a tool you use to manage the information your computer stores in disk files. DOS is a collection of programs that form a foundation for you and your programs to work effectively with your computer. DOS manages the way your monitor and printer display the data you enter into the computer, translates the things you type into codes the computer can understand, and organizes and maintains data on floppy and hard disks.

This book is an introduction to the most widely used Disk Operating System in the world. The material presented here is written in a unique manner—in a learn-by-doing approach that allows you to try new concepts and commands as they are introduced rather than reading through pages of dry, technical material.

What Does This Book Contain?

Learning DOS empowers you to be a better manager of your computer-generated data and, ultimately, a smarter computer user. Designed to help you understand the basics of how the computer works and how you work with the computer, this text presents basic DOS commands that every computer user needs to know and be able to use. This book is not a complete DOS reference and, as such, does not cover every DOS command or every possible permutation of the DOS commands presented. Your computer should have come with a DOS manual. After you finish *MS-DOS SmartStart*, you will be able to use your DOS manual, as well as other DOS reference tools, to learn more complex operations.

The most important thing to remember is that you can learn *only* by doing the exercises in the chapters. No one would argue that you cannot read a book about driving a manual transmission car and expect to be able to drive one without practice. The same is true of the computer!

Chapter 1 describes the components of personal computer systems: the display, the keyboard, the system unit, and peripherals. The last part of the chapter traces the way computers handle data.

Chapter 2 covers the fundamental concepts of how an operating system works, including its different parts. You also learn how to start the computer and begin to explore the two operating modes of DOS: the command line and the DOS Shell.

Chapter 3 is where you really get started. This chapter discusses the basic building blocks of DOS commands, covers the six most frequently used DOS commands, and shows you how to read your DOS manual.

Chapter 4 focuses on file management—using directories and paths—which is critical to working with large hard drives, as most of us now do.

Chapter 5 discusses file maintenance and good computing habits.

Chapter 6 gives a brief listing of advanced commands and simplified descriptions of their functions and uses.

Chapter 7, the final chapter, introduces methods by which you can customize your computing environment.

Appendix A covers DOS installation for both floppy disk and hard disk systems.

Appendix B provides a listing of DOS error messages and instructions for overcoming the impasse they often cause.

Finally, a detailed index helps you quickly find the information you need on a specific topic.

What Hardware Do You Need To Run MS-DOS?

Although MS-DOS can run under a wide variety of set-ups, the minimum requirements necessary to run MS-DOS 5 are at least 256 kilobytes (256K) of system random-access memory (RAM), at least one floppy disk drive, a display (screen/monitor), and a keyboard. These suggestions are minimal; most MS-DOS PCs sold today exceed these requirements.

You cannot use MS-DOS on most computers made by Apple Computer, Inc.; Commodore (except the new Amiga computers, when equipped with additional hardware); or Atari. These computers use operating systems that are sometimes referred to as DOS, but their operating systems are not MS-DOS compatible.

Examples in this book are based on a system with 640K of RAM, one 1.44M floppy disk drive (A:), one 80M hard drive (C:), and a monochrome VGA display.

Conventions Used in This Book

Certain conventions are used throughout *MS-DOS SmartStart* to help you better understand the book's subject.

This book uses a symbolic form to describe command syntax. When you enter a command, you substitute real values for the symbolic name. Examples present commands that you can enter exactly as shown.

DOS commands can have various forms that are correct. The syntax for the DIR command, for example, looks like this if you use symbolic names:

> DIR *d:filename.ext* /W/P

DIR is the command name. The *d:filename.ext* is a symbolic example of a disk drive name and a file name. A real command has actual names rather than symbols.

Some parts of a command are mandatory—information required by MS-DOS. Other command parts are optional. For the preceding DIR command example, only the DIR is mandatory. The rest of the command, *d:filename /W/P*, is optional. When you enter only the mandatory command elements, DOS in many cases uses pre-established values for the optional parts.

You can type upper- or lowercase letters in commands. DOS reads both as uppercase letters. You must type syntax samples shown in this book *letter for letter*, but you can ignore case. Items shown in lowercase letters are variables. You type the appropriate information for the items shown in lowercase letters.

In the example, the lowercase *d:* identifies the disk drive the command will use for its action. Replace the *d:* with *A:* or *C:*. The *filename.ext* stands for the name of a file, including its extension.

Spaces separate some parts of the command line. The slash separates other parts. The separators, or delimiters, are important to DOS because they help DOS break the command apart. Typing *DIR A:* is correct, for example; *DIRA:* is not.

The keys you press and text you type appear in **boldfaced blue** type in the numbered steps and in **boldfaced** type elsewhere. Key combinations are joined by a plus sign: ⇧Shift + F5 in numbered steps and ⇧Shift + F5 elsewhere.

DOS commands, file names, and directory names are written in all capital letters.

On-screen prompts and messages are in a `special typeface`.

Understanding Computer Technology

Have you ever wondered why our society has emotional attachments for some machines but not others? We love our automobiles, for example, but we don't feel the same way about our refrigerators or lawn mowers. This attachment to a car comes from our capacity to develop an interactive relationship with it.

We are social creatures, and that give-and-take relationship may be the reason we have such love affairs with machines like cars and computers. Does a computer take you from zero to sixty in six seconds? Even faster, a computer takes you close to the speed of light. Computers may not turn you into an Indy car driver, but they enable you to participate in a drama of your own making.

1

For some people, a personal computer provides an abundance of emotional "bang for the buck." Others, however, may feel disappointed with their first computer, because it appears to be little more than a big, dumb box. If not for the fact that most people must use a personal computer for school and work, many would prefer to buy a new television set.

Although you may be anxious about getting started, after you understand the basics of how the computer works, you will likely feel much better about learning to use it. This chapter provides a quick but important lesson about the way personal computers work. Learning about the personal computer, its components, and the way the parts work together is like learning about new car controls. You learn about displays and keyboards (dashboards), the CPU (engine), peripherals (tires), and disk drives (the cassette player).

Just as cars have their own terminology, so does the personal computer. These terms are no more mysterious than such terms as *dashboard*, *acceleration*, *mileage*, *pause*, or *reverse*. This chapter explains the components that have become standard for the IBM PC and compatibles. You need not remember every term you read; just try to absorb the big picture.

Objectives

1. To Differentiate between Hardware and Software
2. To Understand the Different Types of Displays
3. To Use the Extended Keyboard Layout
4. To Understand the Relationship among Computer Components
5. To Understand the Basics of Data Processing

Key Terms in This Chapter	
CPU	Central processing unit, the processor in which the actual computing takes place. All problem solving happens here.
Display	The screen or monitor.
Peripheral	Any device, aside from the computer itself, that enables you to perform a task or displays the results (a printer, for example).
Disk	A plastic or metal platter coated with magnetic material, used to store files.
Modem	A device for exchanging data between computers over standard telephone lines.
Input	Any data given to a computer.
Output	Any data transmitted by a computer.
Bit	A *binary digit*, the smallest discrete representation of a value a computer can manipulate. (A computer works with numbers only.)
Byte	A collection of eight bits that a computer usually stores and manipulates as a full character (letter, number, or symbol).
K (kilobyte)	1,024 bytes, used to show size or capacity in computer systems. Technically, the term *kilo* refers to 1,000, but because a base system of 8 is used, a kilobyte has 24 extra bits.
M (megabyte)	1,024 kilobytes.
Data	A broad term referring to words, numbers, symbols, graphics, or sounds. Data is typically the raw facts or materials you enter into the computer.
Information	Data processed by the computer and presented to the user in some meaningful way.
File	A named group of data in electronic form. In word processing, a file can be a letter to a friend. In a database system, a file can be a name and address listing.

1

A Bit of History...

Until the early 1980s, computers were large, expensive machines generally unavailable to individual users. Although the rich could afford them, not many people wanted to fill three rooms of their homes with energy-guzzling machines that served no practical purpose.

Advances in computer technology led to the engineering of smaller computer parts called *integrated circuits*, more commonly known as *chips*. The actual processing in a personal computer takes place on a chip called the *microprocessor*. The newer high-capacity chips save space and energy. The ultimate product of chip technology is the *microcomputer*, as exemplified by your personal computer.

In 1981, International Business Machines (IBM) introduced the IBM Personal Computer, or *PC*. Whether the PC is the best microcomputer is arguable, but IBM did give personal computers respectability in the business community. As a leader in the large-business computer market, IBM held an excellent marketing position. IBM's name, sales force, and corporate contacts made the PC today's standard in home and business computing, creating a market for PCs and a standard on which other firms have built.

Today, many manufacturers produce computers that are in many ways superior to the IBM product line. Rapid technical developments in new companies are raising microcomputer technology to new heights. Even the venerable IBM uses much of this technology in its newest PCs.

Objective 1: To Differentiate between Hardware and Software

A computer system is composed of hardware parts that exist in a variety of configurations (see fig. 1.1). All IBM-compatible computers, however, operate in essentially the same manner. Human needs and preferences more than anything else have determined the size of the computer. The standard model is large enough to contain disk drives and other devices. The portable, on the other hand, is small and light, perfect for computing while on the road.

1

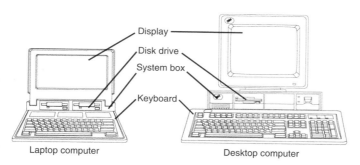

Display
Disk drive
System box
Keyboard

Laptop computer

Desktop computer

Fig. 1.1
Two basic
computer models.

Personal computer systems based on the IBM PC are functionally the same, despite the wide variety of configurations available. Provided that you have the main components, the shape and size of your computer matter very little.

Hardware and software make up the two main segments of a computer system, and you must have both for the computer to work. The purpose of this book is to provide a simple description of the different computer components and their roles in computation, without describing their complexities in detail. If you want to learn more about how computers work, ask your instructor to refer you to several good texts.

Hardware refers to the physical machine and its peripherals—electronics and moving parts of metal and plastic. A VCR, television, tape deck, CD, and turntable also are everyday examples of hardware. In general, if you can touch it, it's hardware.

Software includes the program you are using and the data files created, stored, and run by your PC. These records are the equivalent of textbooks, novels, newspapers, and videotapes. Table 1.1 illustrates the variety of software available for a computer.

The *operating system* provides the working base for all other programs by creating a uniform means for programs to gain access to the full resources of the hardware. Operating systems that help programs to access disks are called *disk operating systems*, or *DOS*.

This book covers the common operating system for IBM PC compatibles: MS-DOS. The IBM versions of DOS and the various versions of Microsoft Corporation's DOS are highly compatible. In fact, they are nearly identical except that IBM calls its version PC DOS and Microsoft calls its version MS-DOS. For this reason, DOS is used in this text as the generic term when referring to both packages.

1

Table 1.1 Computer Software	
Type of Software	*Examples*
Operating systems	MS-DOS (see Chapter 2), OS/2, UNIX
Databases	dBASE IV, Paradox, PC-FILE
Spreadsheets	Lotus 1-2-3, Excel, Quattro Pro
Word processing programs	WordPerfect, Microsoft Word, PC-WRITE
Utilities	Fastback Plus, PC Tools Deluxe, SideKick
Graphics	Harvard Graphics, CorelDRAW!, Lotus Freelance
Integrated programs	Symphony, Microsoft Works, Q&A
Games	Flight Simulator, Tetris, SimCity
Home finance	Quicken, Managing Your Money
Desktop publishing	First Publisher, Ventura Publisher, PageMaker

Objective 2: To Understand the Different Types of Displays

The *video display* (also called the *monitor* or *screen*) is the part of the computer's hardware that produces visual images. To date, the cathode ray tube (CRT) monitor, which operates on the same principle as a television set, provides the crispest, most easily read image. Figure 1.2 shows a screen from a CRT monitor.

Regardless of the display type, all computer screens take electrical signals and translate them into patterns of tiny dots, or *pixels*. (*Pixel* is an acronym coined from the phrase *picture element*.) You can recognize pixels as characters or figures. The more pixels a display contains, the sharper the visual image. The number of pixels in the image determines its *resolution*. As figure 1.3 shows, the higher-resolution image (left), which uses four times as many pixels as the low-resolution image (right), is of much better quality.

```
C:\DATA>DIR/A:D

    Volume in drive C is STACVOL_DSK
    Directory of C:\DATA

    .          <DIR>     03-23-93   2:34a
    ..         <DIR>     03-23-93   2:34a
    TAXES      <DIR>     03-23-93   2:34a
    PERSONAL   <DIR>     03-23-93   2:35a
    HOMEFIN    <DIR>     03-23-93   2:35a
    WORK       <DIR>     03-23-93   2:35a
    NLETTER    <DIR>     03-23-93   2:35a
    DOCS       <DIR>     03-23-93   2:35a
        8 file(s)              0 bytes
                      160866304 bytes free

C:\DATA>
```

Fig. 1.2
A typical CRT
screen.

Fig. 1.3
Comparing high-
and low-resolution
images.

The resolution of the visual image is a function of the display and the *display adapter*, which controls the computer display. In some PCs, the display circuitry is a part of the *motherboard* (see following sections in this chapter). The display adapter can also reside on a separate board that fits into a slot in the computer. The display adapter can be a monochrome display adapter (MDA), Hercules monochrome graphics adapter (MGA), color graphics adapter (CGA), enhanced graphics adapter (EGA), video graphics array adapter (VGA), extended graphics array adapter (XGA), or some other less common display adapter.

Personal computers are interactive. That is, the PC reacts to any action you take, showing the results on the computer display screen. The video display is the normal, or default, location the computer uses to communicate with you.

Table 1.2 lists the most common display types, showing the maximum resolution and the colors available with each type of display adapter.

1

Table 1.2 Resolution and Colors for Display Adapters			
Adapter Type	Graphics Mode	Pixel Resolution	Colors Available
CGA	Medium resolution	320 × 200	4
CGA	High resolution	640 × 20	2
EGA	CGA high resolution	640 × 200	16
EGA	EGA high resolution	640 × 350	16
MGA	Monochrome graphics	720 × 348	2
MDA	Text characters only	80 × 25	2
VGA	Monochrome	640 × 480	2
VGA	VGA high resolution	640 × 480	16
VGA	VGA medium resolution	320 × 200	256
Super VGA	Super VGA	800 × 600	256
Super VGA	1024 Super VGA	1024 × 768	256
XGA	Standard mode	1024 × 768	256
XGA	16-bit color mode	640 × 480	65,536

Objective 3: To Use the Extended Keyboard Layout

The keyboard is the most basic and most common way in which you enter information into the computer, which in turn converts each character you type into code the machine can understand. The keyboard is therefore an *input device*.

Like a typewriter, a computer keyboard contains all the letters of the alphabet, with virtually the same layout for letters, numbers, symbols, and punctuation characters. A computer keyboard differs from a typewriter keyboard in several important ways, however.

The most notable differences are the extra keys that do not appear on a typewriter (see fig. 1.4). Some of these keys may have different functions when used with different programs. Depending on the type of computer and keyboard you use, you also have 10 or 12 special function keys. The standard extended keyboard offers F1 through F12 located across the top of the keyboard.

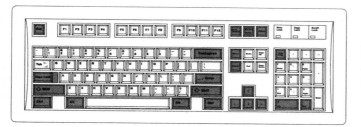

Fig. 1.4
An extended
keyboard.

The *function keys* (located across the top of the keyboard) are shortcuts or command keys. Not all programs use these keys, and some programs use only a few. When used, however, these keys automatically carry out certain operations for you. Programs often use F1, for example, to provide on-line help that displays instructions for a particular operation. The DOS 5 Shell uses F3 to cancel the Shell and F10 to activate the menu.

Table 1.3 describes the standard functions of other special keys on a computer keyboard.

Key	Key Name	Function
Table 1.3 Special Keys on the Computer Keyboard		
↵Enter	Enter	Signals the computer to respond to the commands you type, and functions as a carriage return in programs that simulate the operation of a typewriter
→ ← ↓ ↑	Cursor keys	Move the cursor to a different location on-screen. Included are the arrow, PgUp, PgDn, Home, and Esc.
←Backspace	Backspace	Moves the cursor backward one space at a time, deleting any character in that space

continues

1

Table 1.3 Continued		
Key	*Key Name*	*Function*
[Del]	Delete	Deletes any character at the cursor location
[Ins]	Insert	Inserts any character at the cursor location
[⇧Shift]	Shift	Capitalizes letters when you hold down [⇧Shift] as you press another letter key. When pressed in combination with another key, [⇧Shift] can change the standard function of that key.
[Caps Lock]	Caps Lock	Enables you to enter all capital letters when the key is in the locked position. [Caps Lock] does not, however, shift numbered keys. To release, press [Caps Lock] again.
[Ctrl]	Control	Changes the standard function of a key when pressed in combination with that key
[Alt]	Alternate	Changes the standard function of a key when pressed in combination with that key
[Esc]	Escape	Enables you to escape from a current operation, to the preceding one in some situations. Sometimes [Esc] has no effect on the current operation.
[Num Lock]	Number Lock	Changes the numeric pad from cursor-movement to numeric-function mode
[PrtSc]	Print Screen	On enhanced keyboards, sends the characters displayed to the printer
[Scroll Lock]	Scroll Lock	Causes the cursor-movement keys to scroll the screen instead of moving the cursor

1

Key	*Key Name*	*Function*
Pause	Pause	On enhanced keyboards, suspends display activity until you press another key
Break	Break	Stops a program in progress from running

Many of the function keys are designed for use in combination with other keys (see table 1.4). Holding down Ctrl as you press PrtSc, for example, causes DOS to print continuously what you type. Pressing Ctrl + PrtSc a second time turns off the printing. Break is actually not a separate key. With some keyboards, pressing Ctrl and Scroll Lock together causes a break. With other keyboards, pressing Ctrl and Pause together causes a break.

Table 1.4 DOS Key Combinations	
Keys	*Function*
Ctrl + S	Freezes the display. Pressing any other key restarts the display.
Ctrl + PrtSc	Sends lines to the screen and to the printer; pressing this sequence a second time turns off this function.
Ctrl + C or Ctrl + Break	Stops the execution of a program
Ctrl + Alt + Del	Restarts MS-DOS (system reset)

Objective 4: To Understand the Relationship among Computer Components

Industry engineers designed the standard desktop PC around a box-shaped cabinet, *the system unit*, that connects to all other parts of the computer. Any devices attached to the box are called *peripherals*. The system unit and the peripherals complete the hardware portion of the computer system.

1

The System Unit

The system unit houses all but a few parts of a PC. Included are various circuit boards, disk drives, a power supply, and even a small speaker.

The system units on today's PCs come in many variations of the original design. Desktop models have become smaller to conserve space. Larger models with room for larger, multiple hard disks and other peripherals often use a floor-standing tower design that requires no desk space. Figure 1.5 shows a system unit with a typical placement of hard and floppy disk drives and the system board, also called the *motherboard*.

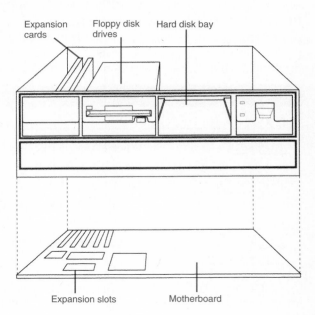

Fig. 1.5
A system unit.

The motherboard holds the main electronic components of the computer, including the microprocessor, the chips that support it, and various other circuits. The motherboard usually contains electrical sockets, called *expansion slots*, into which you can plug various adapter boards, such as a video board.

The chips that provide the computer with its memory are located on the motherboard. You can plug additional memory adapter cards into available expansion slots to increase the system's memory. The number of available expansion slots varies with each PC manufacturer. Most motherboards also have a socket for a math coprocessor, which helps speed up programs that manipulate large volumes of graphics or math equations. Spreadsheet programs and desktop publishing software, for example, benefit from the addition of a math coprocessor chip.

Disk Drives and Disks

Disk drives are complex mechanisms that carry out a fairly simple function: they rotate *disks*, circular platters or pieces of plastic with magnetized surfaces. As the disk rotates, the drive converts electrical signals from the computer and places the information on or retrieves it from magnetic fields on the disk. The storage process is referred to as *writing* data to disk. Disk drives also recover, or *read*, magnetically stored data and present it to the computer as electrical signals. The stored data is not lost when you turn off the computer.

The components of a disk drive are similar to those of a phonograph or CD player. The disk, like a record, rotates. A positioner arm, like a tone arm, moves across the radius of the disk. A head, like a pickup cartridge, translates information into electrical signals. The disk surface, however, does not have spiral grooves; instead, it is recorded in magnetic, concentric rings, or *tracks*. The tighter these tracks are packed on the disk, the greater the storage capacity of the disk.

Two types of disks are available in a variety of data storage capacities. *Floppy disks* are removable, flexible, slower, and a lower capacity; *hard disks*, also called *fixed disks*, are usually high-capacity rigid platters that remain in the system unit.

Hard disks are sealed inside the hard disk drive. Floppy disks are encased in flexible 5 1/4-inch jackets or in rigid 3 1/2-inch jackets (see fig. 1.6). The 3 1/2-inch disks are now the standard floppy disk because they are more durable and can hold more information in less space.

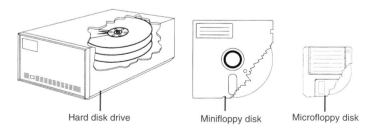

Hard disk drive   Minifloppy disk   Microfloppy disk

Fig. 1.6 Different types of disks.

1

When a computer writes to the disk, it stores groups of data that the operating system identifies as *files*. You can tell that a drive is reading or writing a floppy disk when the small light on the front of the disk drive glows.

Warning: Never open a drive door or eject a disk until the light goes out unless the computer specifically instructs you to do so.

Floppy Disks

Floppy disks store from 360K to 2.88M bytes of data and come in two common sizes. Originally, the 5 1/4-inch floppy disks were called *minifloppies* to distinguish them from the 8-inch disks used on very early personal computers. The 3 1/2-inch disks are sometimes called *microfloppies*. The measurement refers to the size of the disk's jacket. Unless size is important, this book simply refers to both disk types as *floppies*.

In almost all cases, the disk drive uses both sides of a disk for encoding information. Thus, the disk drives and the floppy disks are called *double-sided*.

A drive can handle only one size disk. You cannot read a 5 1/4-inch floppy disk in a 3 1/2-inch disk drive or vice versa. Table 1.5 shows the most common floppy disk capacities in kilobytes (K) or megabytes (M).

Table 1.5 Common Floppy Disk Types	
Disk Type	*Capacity*
5 1/4-inch	
Double density	360K
High density	1.2M
3 1/2-inch	
Double density	720K
High density	1.44M

Make sure that you know your drive's specification before you buy or interchange floppies. Floppies of the same size but with different capacities can be incompatible with a particular disk drive. A high-density disk drive, for example, can format, read, and write to high-density and double-density floppy disks. A double-density disk drive can use only double-density disks.

Because floppies are portable, certain precautions should be taken to protect them. Always store floppies away from sunlight and heat. Keep them away from stereo speakers and telephones (The magnets in these devices can erase part or all of the disk.), cigarette smoke, and anything liquid.

Hard Disks

Hard disks often consist of rigid, multiple-disk platters, with a separate head for each side of each platter. The platters spin at 3,600 RPM, much faster than a floppy disk drive. As the platters spin, the head positioners make small, precise movements above the tracks of the disk. Because of this precision, hard disks can store large quantities of data, from ten to hundreds of megabytes. Hard disks are reasonably rugged devices, and factory sealing prevents contamination of the housing. With proper care, hard disks can deliver years of trouble-free service.

Peripherals

Besides the display, keyboard, and disk drives, a variety of peripherals can be useful to you. Peripherals such as a mouse, printer, modem, joystick, and digitizer enable you to communicate easily with your computer. Using a mouse, for example, with a modern computer program such as a desktop publishing package enables you to take the greatest advantage of the program's features.

The Mouse

The mouse is a pointing device that moves on the surface of your work space and causes the computer to correlate this movement to a cursor, or pointer, on the display. The mouse is shaped to fit comfortably under your hand and has two or three buttons that rest beneath your fingers. The contour and the cable that trails from the unit give the vague appearance of a mouse sitting on the table (see fig. 1.7).

Fig. 1.7
A mouse.

1

The mouse moves a *mouse cursor* (or *pointer*) to make menu selections, draw with graphics programs, move the keyboard cursor swiftly, and select sections of your work to manipulate. Not all software supports a mouse, but many popular programs do. Although mouse functions in different programs can vary, the standard mouse techniques are based on Microsoft Windows. These mouse techniques are also found in the DOS Shell.

Printers

Printers accept signals (*input*) from the CPU and convert the signals to characters (*output*), usually printed on paper. You can classify printers in the following ways:

- By the way they receive input from the computer
- By the way they produce output

You connect printers to the system unit through a *port*, an electrical doorway through which data flows between the system unit and an outside peripheral. A port has its own expansion adapter or shares an expansion adapter with other ports or circuits, such as a multifunction card.

All printers have the job of putting their output on paper. Often this output is text, but it also may be graphics images. Printers fall into three major classifications: *dot-matrix*, *laser*, and *inkjet*. Each printer type produces characters in unique ways. Printers usually are rated by their printing speed and the quality of the finished print. Some printers print by using all the addressable points on-screen, much as a graphics display adapter does. Some printers even produce color prints.

The most common printer, the dot-matrix, uses a print head that contains a row of pins or wires. A motor moves the print head horizontally across the paper. As the print head moves, a slice of each character forms as the printer's controlling circuits fire the proper pins. The wires press against the ribbon and so against the paper, leaving an inked-dot impression. After several tiny horizontal steps, the print head leaves the dot image of a complete character. The process continues for each character on the line. Dot-matrix printers are inexpensive and produce a print quality close to that of a typewriter, and they are commonly used for internal reports. Figure 1.8 shows a typical dot-matrix printer.

Laser printers use a technology that closely resembles that of photocopying. Instead of a light-sensitive drum picking up the image of an original, the drum is painted with the light of a laser diode. The image on the drum transfers to the paper in a high-dot-density output that produces fully formed printed characters. Laser printers can also produce graphics. The high-quality text and graphics combination is useful for desktop publishing as well as general business correspondence.

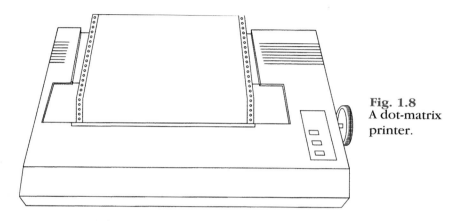

Fig. 1.8
A dot-matrix
printer.

The inkjet printer literally sprays words and graphics on a page in near silence. Moderately priced, the print quality rivals that of a laser printer. Of all the printers, only a laser is faster and sharper.

Modems

A *modem* is a serial communications peripheral that helps your PC communicate with other computers over standard telephone lines, sending or receiving characters or data one bit at a time (see fig. 1.9). Most modems communicate with other modems at speeds from 300 to 9,600 *bps*, or bits per second. The most common speed for modems today is 2,400 bps. Modems need special communications software to coordinate data exchanges with other modems. A modem can be installed internally in the system unit or reside in a small case outside the computer.

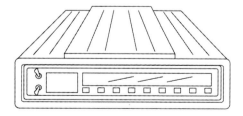

Fig. 1.9
A modem.

1

You use a modem to send or receive files to another computer, to use a computerized *bulletin board system* (BBS), or to access an on-line service such as Prodigy or CompuServe.

Objective 5: To Understand the Basics of Data Processing

Now that you have learned about the essential parts of the computer system, you are ready for an overview of how all these parts carry out the job of computing. Fortunately, you do not have to know the details of a computer's operation to produce finished work. If you explore just a little, however, you will adjust more quickly to using your computer.

Computers perform many useful tasks by accepting data as input, processing it, and releasing it as output. The data can be any information, such as a set of numbers, a memo, or an arrow key that moves a game symbol.

The computer translates input into electrical signals that move through a set of electronic controls. You can think of output in four ways:

- As characters the computer displays on-screen
- As signals the computer holds in its memory
- As codes stored magnetically on disk
- As permanent images and graphics printed on paper

Computers receive and send output in the form of electrical signals that are stable in two states: on and off. You can think of these states as you think of electricity flowing to a light switch that you can turn on and off. Computers contain millions of electronic switches that can be on or off. All input and output follows this two-state principle.

Binary, the computer term for the two-state principle, consists of signals that make up true computer language. Computers interpret data as two binary digits, or *bits*—0 and 1. For convenience, computers group eight bits together. This eight-bit grouping, or *byte*, is sometimes packaged in two-, four-, or eight-byte packages when the computer moves information internally.

Computers move bits and bytes across electrical highways called *buses*. Normally, the computer contains three buses: the *data bus*, the *control bus*, and the *address bus*. The microprocessor connects to all three buses and supervises their activity. The CPU uses the data bus to determine *what* the data should be, the control bus to confirm *how* the electrical operations should proceed, and the address bus to determine *where* the data is to be positioned in memory.

Because the microprocessor can call on this memory at any address, and in any order, it is called *random-access memory*, or *RAM*. The CPU reads and executes program instructions held in RAM, and the resulting computations are stored in RAM.

Some computer information is permanent. This permanent memory, called *read-only memory* (or *ROM*), is useful for holding unalterable instructions in the computer system.

The microprocessor depends on you to give it instructions in the form of a *program*, a set of binary-coded instructions that produce a desired result. The microprocessor decodes the binary information and carries out the instruction from the program. For example, figure 1.10 diagrams the complex series of steps the computer takes to display a word on-screen.

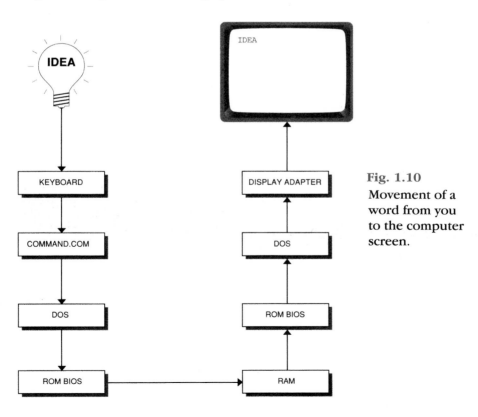

Fig. 1.10
Movement of a word from you to the computer screen.

1

You can start from scratch and type programs or data into the computer every time you turn on the power. Of course, you don't want to do this if you don't have to. Luckily, the computer stores both instructions and start-up data, usually in binary form in files on disk. To the computer, a file is just a collection of bytes identified by a unique name. These bytes can be a memo, a word processing program, or some other program. A file's function is to hold binary data or programs safely until you type a command on the keyboard to direct the microprocessor to call for that data or program file. When the call comes, the drive reads the file and writes its contents into RAM.

Chapter Summary

This chapter introduces the basic vocabulary and concept set that you need to use an IBM-compatible personal computer. The text describes the basic components and the way they interact with each other, giving you a brief but accurate overview of the data processing process that occurs inside the computer. Now you are ready to begin working with your computer and DOS.

Testing Your Knowledge

True/False Questions

1. Computer components can be grouped into two basic categories: hardware and software.
2. Operating systems enable you to type papers and prepare budgets.
3. Hard disks typically have much greater storage capacities than floppy disks.
4. The control bus, data bus, and address bus are the pathways over which electrical signals are sent inside the computer.
5. The computer operating system (DOS) is responsible for directing input to the appropriate place and directing output to the appropriate place.

Multiple Choice Questions

1

1. The system box, display, keyboard, and disk drive are all examples of
   - A. input devices.
   - B. output devices.
   - C. hardware.
   - D. software.
   - E. none of the above

2. Electrical sockets into which computer peripheral control devices can be placed are called
   - A. plug-ins.
   - B. motherboard.
   - C. banks.
   - D. expansion slots.
   - E. none of the above

3. Which of the following devices is used in computer-to-computer communications over standard phone lines?
   - A. printer
   - B. CPU
   - C. RAM
   - D. modem
   - E. none of the above

4. Which of the following devices holds unalterable, permanent information?
   - A. RAM
   - B. ROM
   - C. hard disks
   - D. floppy disks
   - E. none of the above

5. A grouping of 8 bits is called a
   - A. byte.
   - B. kilobyte.
   - C. megabyte.
   - D. gigabyte.
   - E. none of the above

1

Fill-in-the-Blank Questions

1. WordPerfect, Lotus, dBASE, and DOS are all examples of
   _____*software*_____.
2. The most common input device is a(n)_____*key board*_____.
3. A _____*Dot - matrix*_____ printer is the most cost-effective, high-quality alternative to a laser printer.
4. _____*eight*_____ bits are required to represent one character.
5. A _____*modem*_____ is a device that enables two computers to communicate with one another.

Review: Short Projects

1. Locating Campus Computer Facilities

   Find a brochure about the computer facilities on your campus and determine how many IBM-compatible computing sites are available for student use and where these sites are located.

2. Locating Non-DOS Operating Systems

   Determine from the brochure what other operating systems besides DOS are in use on your campus.

3. Recommending Operating Systems

   Which of the operating systems you investigated in Short Project 2 would you recommend be expanded for your school? Why?

Review: Long Projects

1. Investigating Software Availability

   Visit a software store in your area, and examine the different types of software available in the different categories. Categories may include word processors, spreadsheets, databases, accounting, drawing, utilities, and games.

   Does each software package specify the type of equipment required? Are minimum RAM requirements mentioned? Disk requirements?

   What does each software product have in common with others in its own category as well as with software in other categories?

2. Selecting a Computer System

   A friend has decided to start a lawn maintenance business and has asked your advice on what software and hardware he will need to run his business efficiently. Design an appropriate computer package for your friend, specifying what computer and computer peripherals will be required, along with recommended software.

   Hint: Because your friend has no computer experience, you might consider software packages with a low learning curve.

1

The Anatomy of DOS

T he *disk operating system* is one of the most important aspects of working with microcomputers because it
determines the way you communicate with your computer. Some disk operating systems have pictures and
menus to command the computer; others enable you to
enter commands directly through a command line. DOS
(pronounced DAHS) provides you with the best of both
worlds.

Chapter 1 discusses software, mechanics, computer
systems, and component parts. This chapter introduces
DOS, the most important link between hardware, other
software, and you. This chapter brings DOS closer to
you by defining the disk operating system and its uses.

Objectives

1. To Understand the Functions of an Operating
   System
2. To Learn the Conventions for Naming DOS Files
3. To Understand How DOS Manages Files and
   Applications
4. To Perform a Cold and Warm Boot
5. To Access and Use the DOS Shell

2

Key Terms in This Chapter	
Program	Instructions that tell a computer what to do and how to do it
BIOS	Basic Input/Output System—the program that performs basic communications between the computer and the peripherals
Applications program	Instructions that tell the computer to perform a program-specific task, such as word processing
Interface	A connection between parts of the computer, particularly between hardware devices; also refers to the interaction between you and an applications program
Command	An instruction that you give to DOS to perform a task
Batch file	A series of DOS commands placed in a disk file. DOS executes batch-file commands one at a time.
Cold boot	Starting your PC by turning on the power switch
Warm boot	Restarting, or resetting, your PC without turning it off, by pressing Ctrl + Alt + Del
Cursor	The blinking line or solid block that marks where the next keyboard entry will appear
Prompt	A symbol, character, or characters that request information before anything else can happen
DOS Shell	A graphical menu-driven interface that enables you to execute DOS commands easily without having to learn the names of commands
Graphical User Interface (GUI)	A graphical way to present information on-screen and accept information from the user (used in the DOS Shell and Microsoft Windows)

Shell	The interface used to operate DOS
Icon	A small picture that graphically represents an object, such as a file or a program
Mouse pointer	A symbol, usually an arrow, that shows the position of the mouse
Click	To press the mouse button
Double-click	To press and release the mouse button twice, quickly
Drag	To press and hold down the mouse button as you move the mouse
Selection cursor	The highlighted band or area that indicates that an item is selected
Pull-down menu	A secondary menu that drops down below the menu bar when you select a menu option
Dialog box	A window that displays options when a command needs additional information

Objective 1: To Understand the Functions of an Operating System

An operating system is a collection of computer programs that provides special services to other programs and to the computer user. You or a program tells DOS what action to take, and DOS directs the hardware to carry out the command. Figure 2.1 provides a visual representation of some of the tasks that DOS performs.

If a computer's operating system did not supply these services, you would have to deal directly with the details of controlling the hardware. Without the disk operating system, for example, every computer program would have to hold instructions telling the hardware each step to take to do its job.

Because operating systems already contain instructions, you or a program can call on DOS to control your computer. The name *disk operating system* comes from the attention given to the disks in the computer system.

2

IBM-compatible personal computers use MS-DOS, the disk operating system developed (although not invented) by Microsoft Corporation. When you read about DOS in this book, you can assume that the information is generalized to cover your manufacturer's version of DOS. In special cases, differences are noted.

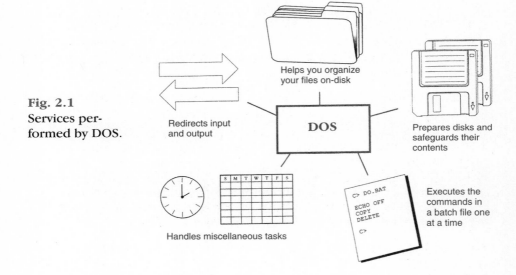

Fig. 2.1
Services per-
formed by DOS.

Helps you organize
your files on-disk

Redirects input
and output

DOS

Prepares disks and
safeguards their
contents

Handles miscellaneous tasks

Executes the
commands in
a batch file one
at a time

DOS has three main functional components:

* The command interpreter
* The file and input/output system
* Utilities

These components are contained in files that come with your DOS package. In the following sections, you are introduced to the components and to their duties.

The Command Interpreter

The *command interpreter* is DOS's "electronic butler"; it interacts with you through the keyboard and screen when you operate your computer. The command interpreter is also known as the command processor and is often referred to as COMMAND.COM.

When you give COMMAND.COM an instruction, it determines what you want to do and starts the program you request. COMMAND.COM also contains the most commonly used commands, such as those to list the files on a disk or copy a file.

You can give commands to COMMAND.COM in two ways. You can type a command at the DOS prompt, or you can use the *DOS Shell*. Both techniques are demonstrated in the next chapter.

When COMMAND.COM displays the *DOS prompt*, you know that COMMAND.COM is ready to receive a command. When you enter a command, you are really telling COMMAND.COM to interpret what you type and to process your input so that DOS can take the appropriate action. A typical DOS prompt looks like this: C:\>

The File and Input/Output System

The file and input/output system is made up of so-called *hidden* files and programs that are actually part of your computer's hardware. Your computer cannot run without these special hidden files. Hidden files do not appear on a normal directory listing or other lists of files, and you cannot delete or copy hidden files. They are not hidden to trick you into thinking that they don't exist but to protect them from accidental deletions or changes.

The hidden files interact with programs stored in special read-only memory (ROM), which is part of your computer's hardware. The special ROM is called the ROM Basic Input/Output System, or *BIOS*. Responding to a program's request for service, the system files translate the request and pass it to the ROM BIOS. The BIOS provides a further translation of the request, linking the request to the hardware.

The Utility Files

The DOS utilities are command programs that are not built into COMMAND.COM. DOS utilities carry out useful housekeeping tasks, such as preparing disks, comparing files, finding free space on a disk, and printing in the background. Several utilities supply statistics on disk size and memory; others compare disks and files.

The utility programs are files that reside on disk and are loaded into memory by COMMAND.COM when you type their command names. These commands are often called *external* commands because they are not built into COMMAND.COM. Commands built into COMMAND.COM are called *internal* commands.

2

By now you may suspect that DOS makes technical moves that are difficult to understand. True, much of DOS's activity is technical, but the features you need to master to make DOS work for you are easy to understand. This section briefly describes the DOS functions you will use repeatedly as your expertise grows. Later chapters treat these topics in more detail.

Objective 2: To Learn the Conventions for Naming DOS Files

The files on your disks may fall into a variety of categories, and the *file name* can help you determine what type the file is. A file name can be as long as eight characters or as short as one character. You can also add a period and a three-character extension to the file name: SADIE.LET, for example. In many cases, you provide the file name, and the program adds the file extension. Over the years, a kind of shorthand has developed to simplify identification of computer files by using the three-character extension at the end of the file name.

Programs are files that contain computer instructions. Most DOS files represent programs. When you buy an *applications program*, you are buying computer instructions to perform certain tasks, such as word processing or spreadsheet manipulation. A program file is called an *executable file* because, in computer lingo, to run a program is to *execute* it. Executable files usually have an EXE file extension.

A *command file* is an executable file in a special format with a COM file extension. Some DOS program files have EXE file extensions, and others have COM file extensions. You can ignore the different extensions and treat both types of files as programs.

Text files, such as a letter to your Aunt Sadie, are created by a word processing or text editing program. Text files usually have a TXT or DOC file extension. Another kind of text file is the one that comes with most software and is often titled README.DOC. This file supplies additional instructions for the program, that were not included in the printed manual.

Many other types of files and file formats exist. Application programs may have their own special formats and file extensions. Table 2.1 shows some examples of the files and file extensions used by DOS.

Table 2.1 Files Contained in DOS	
File Name	*Description*
COMMAND.COM, FORMAT.COM, EDIT.COM, MIRROR.COM	The COM file extension identifies a command file.
EGA.CPI, LCD.CPI	Files with CPI extensions operate the display screen.
CATAPULT.BAS	Identifies a program written in the BASIC language. Many games are written in BASIC.
AUTOEXEC.BAT	A batch file. DOS looks for this batch file and runs it automatically when you start your computer.
BACKUP.EXE, MEM.EXE, RESTORE.EXE, CHKDSK.EXE DOSSHELL.HLP, EDIT.HLP	Executable program files. HLP files display on-screen assistance.
DOSSHELL.INI	Initiation files that contain program default information
KEYBOARD.SYS, CONFIG.SYS, ANSI.SYS	System files. They are also called device drivers.

2

Note: If you ask DOS for a listing of the files contained on a disk, the dot for the extension does not appear. Table 2.2 compares the listing appearance with the real file name.

Objective 3: To Understand How DOS Manages Files and Applications

DOS's many activities can be organized into several general categories. The following sections describe the most frequently performed DOS services.

2

Table 2.2 File Name Listing	
On-Screen Listing	*Real File Name*
COMMAND COM	COMMAND.COM
EGA CPI	EGA.CPI
AUTOEXEC BAT	AUTOEXEC.BAT
FIND EXE	FIND.EXE
SELECT HLP	SELECT.HLP
DOSSHELL INI	DOSSHELL.INI
KEYBOARD SYS	KEYBOARD.SYS

Managing Files

One of DOS's main functions is to help you organize the files you store on your disks. Organized files are a sign of good computer housekeeping, which becomes crucial as you take advantage of the storage capacity available on your disks.

The smallest capacity floppy disk can hold the equivalent of 100 letter-sized pages of information. If each page of information makes up one file, you have 100 files to track. If you use disks that hold more information than a standard floppy (such as a hard disk), file organization becomes even more important.

Fortunately, DOS gives you the tools to be a good computer housekeeper. DOS lists files, tells their names and sizes, and gives the dates they were created or last modified. You can use this information for many organizational purposes. In addition to organizing files, DOS provides commands to dupli-cate files, discard outdated files, and replace files whose file names match.

Managing Disks

Certain DOS functions are essential to all computer users. For example, all disks must be prepared, or *formatted*, before they can be used in your computer. (You learn to format a disk in Chapter 3.) Other disk-management functions that DOS performs include the following:

- Labeling disks electronically
- Making reconstructible backup copies of files for security purposes
- Restoring damaged files
- Copying disks
- Checking a disk for errors

Running Applications Programs

Computers require complex, exact instructions, or *programs*, to provide you with useful output. Computing would be totally impractical if you had to write a program for each job you needed to complete. Happily, this extra work is not necessary. Programmers spend months writing the specialized code that permits computers to function as many different tools: word processor, database manipulator, spreadsheet, and graphics generator. Through these programs, called *applications programs*, the computer's capabilities are applied to a task.

Applications programs are distributed on disks. DOS acts as a go-between, enabling you to access these programs through the computer. By inserting a disk into a computer's disk drive and pressing a few keys on the keyboard, you have at hand a wide variety of applications.

Applications programs constantly read data from disk files to review what you have typed and to send information to the screen or printer. These input and output operations are common repetitive computer tasks. DOS furnishes applications with a simple connection, or program *interface*, that sees to the details of these repetitive activities. As you use your computer, you may want to view information about disk files, memory size, and computer configuration.

Knowing the Uses of DOS

Because a computer cannot start or operate without DOS, anyone using a personal computer can benefit from a working knowledge of DOS. If you take the time to learn DOS basics, you will become much more skilled at computing.

Objective 4: To Perform a Cold and Warm Boot

2

Most computer terms have simple origins and meanings. To *boot* your PC, for example, means to turn on the computer (a *cold boot*) or to instruct the computer to reset itself without your turning it off (a *warm boot*). The derivation comes from the expression "pull yourself up by your bootstraps."

With early computers, operators started by entering a binary program, called a *bootstrap loader*, and instructing the computer to run the program. The term *booting* stuck; even now, it is still used to refer to the start-up procedure, which with today's DOS, is an easy process.

Most computers purchased today come equipped with a hard disk permanently fixed inside the computer; a hard disk can hold hundreds of times more data than a floppy disk. The operating system (DOS) is stored on your computer's hard disk. This book provides practice booting from a hard disk. If your computer has two floppy drives and no hard disk, ask your instructor for special instructions for booting your computer.

The Cold Boot

A *cold boot* is the procedure used to start the computer when it has been physically turned off. The term *cold* is used because no electricity is flowing to the computer to keep it warm.

A cold boot is simply a matter of flipping switches. You need to turn on your monitor first and then your system unit. Different manufacturers place the power switches in different places, so look around your system unit and monitor to find the on-off switches. Many computers now have Reset buttons. The Reset button functions the same way as flipping the system unit and monitor power switches off and on; use the Reset button instead of the actual power switches if the button is available.

Exercise 4.1: Performing a Cold Boot

To start your computer when it is turned off, follow these steps:

1. Locate the on-off switch for your monitor.
2. Turn on the power to the monitor.
3. Locate the on-off switch for your PC, and turn on the power.
4. Now sit back and watch.

If your computer has a Reset button, turn on the computer, and then press Reset. Your system will turn off and back on again as you watch.

When the computer is turned off, all the memory except ROM is blank. As soon as electricity flows to the computer, ROM becomes active and begins to direct traffic. ROM is the storage place for the *boot instruction set*, a set of instructions that tell the computer how to boot. After the boot instruction set is loaded, ROM searches for the disk operating system (DOS, in this case). When you see the light on your hard drive come on the first time, ROM is looking for DOS. After finding DOS, ROM tells DOS to take over and finish the boot procedure.

DOS begins by checking the RAM in the computer to see how much it has to work with. When DOS is finished checking the RAM, DOS displays the amount available. DOS also checks to see how many disk drives the computer has, whether a monitor is connected and turned on, whether the keyboard and mouse are working, and whether a printer is connected to the system unit. DOS then runs a file called AUTOEXEC.BAT, which contains final instructions on the way DOS is to behave, and soon the DOS prompt or DOS Shell appears. You are now officially booted.

The Warm Boot

A *warm boot* occurs when you restart a computer that is already on. You may need to reboot using a warm boot for many reasons. The most common reason for rebooting is that the computer occasionally locks up, and the only way to reset the system is to start over.

Note: A warm boot is much better for the computer system than a cold boot; it causes less wear and tear on the power switch.

Look at the keyboard and locate the Ctrl, Alt, and Del keys. The warm boot requires simultaneously pressing and holding Ctrl and Alt and then pressing Del. The PC skips the preliminary tests and immediately loads DOS. Don't worry if nothing happens on the first try. With some systems that have run programs, you may have to use Ctrl + Alt + Del twice.

Exercise 4.2: Performing a Warm Boot

In this exercise, you reboot your computer using the warm boot method. Follow these steps:

1. Eject any disks from the floppy disk drives, if necessary.

2. Hold down Ctrl and Alt with your left hand.
3. Press Del with your right hand.
4. Release all three keys.

    The system reboots.

Objective 5: To Access and Use the DOS Shell

With the advent of DOS 4.0, users had the choice of viewing DOS in the traditional way, with a prompt and command-line interface, or through a new, more visual method known as the *DOS Shell*. The DOS Shell in DOS 4.0 wasn't very good, and most users chose to use the familiar command line instead. The much improved Shell interface for DOS 5 warrants a second look, however.

The remainder of this chapter discusses the DOS Shell, explains what individual portions of the screen mean, and illustrates some basic navigational methods for performing tasks in the Shell. This chapter provides the only major lesson devoted to the DOS Shell in this book, providing sufficient information so that you can use the Shell on your own.

The *prompt view* is the traditional, simple look of DOS. The DOS prompt appears on a plain screen, telling you that DOS is waiting for a command. If you do not have a hard disk, the standard prompt is one letter of the alphabet, which represents the current (or active or logged) drive, followed by a *greater than* (>) symbol. If you boot your system with the DOS disk in drive A, A> appears on-screen.

If your computer has a hard disk drive, the standard DOS prompt is the drive letter, followed by a colon (:), followed by the directory path, followed by the greater than symbol. After you boot from the hard disk, C:\> appears on-screen.

You may be wondering why the letter B is missing. The first floppy drive is always called drive A. The second floppy drive, if present, is always called drive B. The first hard drive, if present, is almost always called drive C. If you have only one floppy drive and a hard drive, the hard drive is called drive C, and no drive B exists.

Looking at the DOS Shell

The *Shell view* provides a full-screen window with menus, pop-up help boxes, and a graphic presentation of directories and files. You can issue standard DOS commands using a mouse or the keyboard to point to and select pull-down menus and dialog boxes. You do not have to remember the names of commands to use the Shell. You just select actions from menus, type answers to questions, and check off options in dialog boxes.

If you have ever used Excel, Windows, or Works, the DOS Shell will be familiar to you, for it is similar to Microsoft's other mouse-and-menu products. Even if you have never used these products, you will find the Shell easy to learn and to use.

A mouse is almost a prerequisite for many shells, desktop publishing programs, and the latest generation of Windows software. On the other hand, if you know your way around a typewriter keyboard, you can perform some functions more quickly in the DOS Shell by using the cursor-movement keys.

The Shell view provides a graphical view of the structure of your computer files. Some users feel that the Shell is easier to use than the DOS command-line interface. However, many DOS commands cannot be executed in the DOS Shell, and many fast typists find that selecting menu items in the Shell is slower than entering the commands at the prompt level with the standard keyboard. The choice is yours.

The Shell provides a visual presentation of DOS, with *action options* from which you make selections. You can manage your computer from the action options, which enable you to use the function and cursor-movement keys to select the same DOS commands you type at the DOS prompt.

The DOS Shell automatically displays information about a disk, a directory, and programs (see fig. 2.2). You use a mouse or the keyboard to move around the display, to select menu items, and to execute programs.

When you make a selection from the menu, the menu commands for that selection pull down. Figure 2.3 shows the File pull-down menu.

To move, copy, and rename files or to execute other commands, you don't have to remember the command names; you just select items from command menus. When you need to supply more information to complete a command, DOS presents you with a *dialog box*. You type the answers to the prompts, and DOS completes the command.

2

Although this book concentrates on using DOS commands at the command-line level, you may want to experiment with the DOS Shell on your own. Again, after you understand how DOS works with the computer, you will have the necessary knowledge to work with DOS in either way.

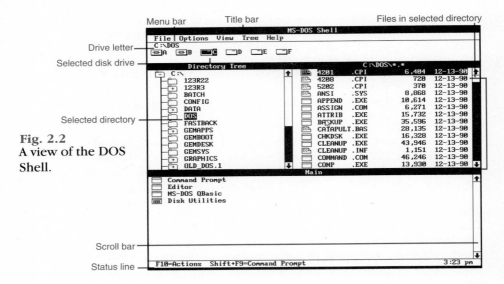

Fig. 2.2
A view of the DOS Shell.

Fig. 2.3
The File pull-down menu.

Exercise 5.1: Examining the DOS Shell

In this exercise, you load the DOS Shell by entering a command at the DOS prompt. You can then examine the Shell to determine what it displays. Follow these steps:

1. Boot your computer, if necessary.
2. To load the Shell from the DOS prompt, type

   dosshell

3. Press ⮐Enter.

   The Shell displays information about your files and programs. You can choose from a number of different ways or views to display this data.

4. Read through the screen descriptions in table 2.3, making sure that you can locate the different components of the DOS Shell view.

When you start the DOS Shell, you see a full-screen display (refer to fig. 2.2). This display contains much information that is displayed automatically, including the list of disk drives in your computer, the files in the root directory, and a list of some of the DOS programs available.

The initial Shell view can contain up to 25 lines of text. Use this display mode if your monitor can display text only or if you have a low-resolution CGA monitor. This view has three areas: the directory tree at the left, the list of files at the right, and a list of programs at the bottom.

A number of items in the Shell display appear highlighted to indicate that these items are selected. One of the disk drive letters, for example, is highlighted to indicate that it is the selected drive and that the directories displayed are located in the selected drive. Likewise, for the directory that is highlighted, the Shell displays the files in that directory. When you use the Shell, knowing which items are selected is important, as you will see throughout this chapter.

Table 2.3 The Parts of the DOS Shell	
Part	*Description*
Title bar	The name of the current window or dialog box
Menu bar	The bar below the main window title bar with the list of pull-down menu options

continues

2

Table 2.3 Continued	
Part	*Description*
Drive letters	The list of disk drives that your computer recognizes. The selected drive is highlighted. Drives A and B are floppy drives, and the first hard disk drive is C. If your hard disk drive or drives are partitioned into separate logical drives, each drive letter is treated as a separate physical drive.
Directory Tree area title	Identifies the Directory Tree area. The title is highlighted when this area is selected.
Directory Tree area	Shows the directories for the selected drive. The selected directory is highlighted.
Files area title	Identifies the files area. The title is highlighted when this area is selected.
Files area	Shows the files for the selected directory. The selected file is highlighted.
Program area title	Identifies the program area. The title is highlighted when this area is selected.
Program area	Lists the programs available from the current program group and lists other program groups.
Selection cursor	A highlighted area or band that indicates the selected drive, directory, file, or program
Status line	The bottom line of the Shell display, which shows some function key actions, messages, and the current time
Mouse pointer	Shows the current position of the mouse on the display. You use the mouse pointer to select items.
Scroll bars	Used to scroll a list of directories, files, or programs that is too long to fit in the display area.

From the Shell, simply by choosing menu options, you can execute DOS commands, run programs, view the contents of files, and change the appearance of the Shell display. You do not have to remember command names or the format and parameters of the commands. Just browse through the Shell to see what commands are available.

Selecting Items

One of the main concepts of the Shell graphical user interface is that of selecting an item and then performing some action with the item. An item can be a disk, directory, file, or program. You choose an item by using the mouse or the keyboard; you choose an action from a menu or through a shortcut key.

Exercise 5.2: Using the Mouse

Although you can use the Shell with the keyboard or the mouse, you will find using the mouse much easier. The graphical user interface was designed for use with a mouse. In fact, the mouse was designed and developed specifically to control a graphical interface.

To practice using the mouse, follow these steps:

1. Boot your computer, if necessary.
2. Load the DOS Shell, if necessary, following the instructions given in Exercise 5.1.
3. When the DOS Shell appears on-screen, move the mouse around. Notice that as you move the mouse, the mouse pointer also moves on-screen in the same direction. The mouse pointer takes on different shapes in the various modes of DOS 5. In Graphics mode, for example, the mouse pointer is an arrow. In Text mode, the pointer is a gray rectangle. The mouse pointer changes shape to indicate the action taking place.
4. To choose an item using the mouse, move the tip of the mouse pointer over the item you want to select; then click the left button. (Your mouse may have two or three buttons, but in most cases you use only the left button.) Use gentle pressure; if you press too hard, you may move the mouse as you click and make the wrong selection.
5. Select the file CONFIG.SYS, located on the C drive, by clicking the name once. Your screen should look like the screen shown in figure 2.4, with CONFIG.SYS highlighted.

2

Fig. 2.4
The file
CONFIG.SYS
selected.

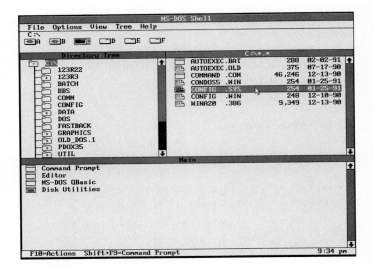

Exercise 5.3: Using the Keyboard

Although slower than using a mouse, the keyboard procedure works just as well. To choose an item using the keyboard, follow these steps:

1. Boot your computer, if necessary.

2. Load the DOS Shell, if necessary, following the instructions given in Exercise 5.1.

3. Press `Tab⇄` to select the area you want. (The selected area title bar is highlighted as you press `Tab⇄`.) If no title bar is highlighted, the selection area is the disk drives area near the top of the display (refer to fig. 2.4).

4. Press `↑` and `↓` to move the selection cursor to the directory, file, or program you want to select. In the disk drives area, press `←` and `→` to choose a different disk drive. Press `⏎Enter`.

Making Menu Selections

After choosing an item, you specify an action, most often from a menu. The initial menu options—**File**, **Options**, **View**, **Tree**, and **Help**—are listed in the menu bar.

When you choose a menu, you activate a pull-down menu. DOS displays a pull-down menu with a list of commands (refer to fig. 2.3). From this menu, you choose the specific action, such as a DOS command, that you want to use. Menu options that appear dimmed are unavailable for selection.

Exercise 5.4: Using the Mouse with Menus

In this exercise, you practice using the mouse to choose a menu option and a command from the pull-down menu:

1. Boot your computer, and load the DOS Shell, if necessary.
2. Move the mouse pointer to the File menu name, and hold down the mouse button. From this point on, this process is referred to as *choosing*. (Choose File, View, for example, means to choose the View option from the File menu). Notice the options that are available from this menu.
3. Examine the options under each of the five pull-down menus.
4. Choose File, Exit to exit from the DOS Shell and return to the DOS prompt.

Exercise 5.5: Using the Keyboard with Menus

To use the keyboard to choose a menu option and a command from the pull-down menu, follow these steps:

1. Boot your computer, and load the DOS Shell, if necessary.
2. Press Alt or F10 to activate the menu bar. This action highlights the File menu option (see fig. 2.5).
3. Press → to highlight Options.
4. Press ↓ to activate the pull-down list box for the Options menu.
5. Press ↑ and ↓ to highlight the various commands in the pull-down menu.

Exercise 5.6: Cancelling Menus

Selecting the wrong menu option is easy, especially when you are first learning how to use the mouse. Exercise 2.6 left you in the Options menu. To cancel your Options selection and move to a different menu, follow these steps:

2

1. Make sure that you are in the Options menu.
2. If you are using a mouse, click anywhere outside the Options menu.
3. If you are using the keyboard, press [Esc] to cancel the selection.

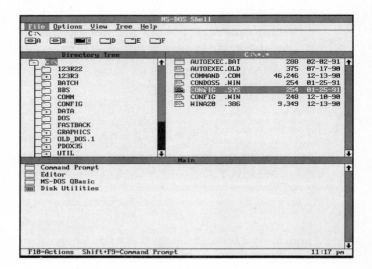

Fig. 2.5
Activating the
menu bar.

Using Dialog Boxes

Some commands need additional information before DOS can carry them out. When DOS needs additional information, it opens a *dialog box* (see fig. 2.6). When a command has a dialog box, the command name in the pull-down list box is followed by an ellipsis (...), as shown in figure 2.7. A dialog box may ask for one or more pieces of information, depending on the command.

The dialog box is one of the most powerful features of a graphical user interface. With other menu systems, you must go through many different levels of menus to get to the one you want. You may even have to make more than 10 selections to execute one command. With pull-down menus and dialog boxes, however, you select a menu and then a command. If additional information is needed, it is requested in a dialog box that may be as large as a full screen if necessary. Figure 2.6 shows the Display Options dialog box.

Dialog boxes may contain the following elements:

- *Text box.* A box in which you type text, such as a file name
- *Check box.* An on/off or yes/no question preceded by a pair of square brackets. If the option is selected, an X appears in the square brackets.

2

- *Option button*. A circle appearing next to a specific option. The selected option has a black dot in the circle; you can choose only one option at a time.
- *List box*. A list of options (like option buttons in a different format). You can choose only one option from the list.
- *Command button*. A representation of an action you can take from the dialog box. OK processes the command, Cancel cancels the command, and Help displays on-line help for the dialog box.

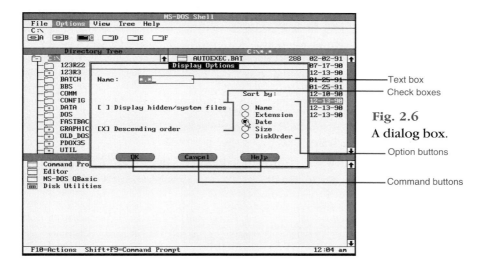

Text box
Check boxes

Fig. 2.6
A dialog box.

Option buttons

Command buttons

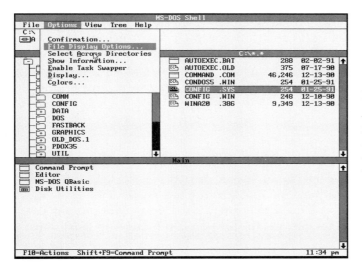

Fig. 2.7
A command followed by an ellipsis to indicate that it has a dialog box.

Exercise 5.7: Entering Text, Selecting Options, and Using Command Buttons in Dialog Boxes

2

In this exercise, you practice using the various components of a dialog box. The skills you learn in this section are necessary for working with any graphical user interface system that uses dialog boxes. Follow these steps:

1. Boot your computer, and start the DOS Shell, if necessary.

2. Choose Options, File Display Options.

3. Move the mouse pointer to the text box, and click. If you are using the keyboard, press Tab⇥ or Ctrl+Tab⇥ to move to the text box.

4. To type over an existing entry, just type the new entry. For this exercise, type *.**bat** to change the view to display batch files only. (These files are discussed elsewhere in this chapter.)

5. To display hidden system files as well, place an X in the selection box by moving the mouse pointer between the square brackets and clicking. If you are using the keyboard, press Tab⇥ or ⇧Shift+Tab⇥ to move to the check box and press the **space bar**.

6. To sort the display by extension so that all the same types of files are displayed together, point the mouse to the circle next to Extension in the Sort By section of the dialog box, and click. If you change your mind, you can click the same circle again to deselect the option.

   When you have set the dialog box according to your needs, you are ready to have DOS execute the commands.

7. Choose the OK command button at the bottom of the dialog box. With the mouse, point to the OK button, and click. With the keyboard, press Tab⇥ or ⇧Shift+Tab⇥ until an underline appears in the command button you want; then press Esc or the **space bar**. You can press Esc to cancel a command at any time.

After making these changes, you should see only the hidden files and BAT files in your view. These files should be organized, or *sorted,* by their extension, in alphabetical order.

Using Scroll Bars

Sometimes a list is too long to fit within a display area or dialog box. If you use a mouse, you can use the *scroll bars* that run the length of the list to view text that is not visible. Inside the scroll bar is a *scroll box*, a gray rectangular box that represents the position and the fraction of the data from the list that is currently displayed. If the scroll box is small compared to the total length of the scroll bar, the list is long, and you can see only a small part of the list at any one time.

In figure 2.8, the scroll box for the Directory Tree area starts at the top of the scroll bar, indicating that the top of the list of directories (for the root C:\ directory) is displayed. The length of the scroll box is about three-quarters the length of the scroll bar, indicating that about three-quarters of the directories in the list are visible.

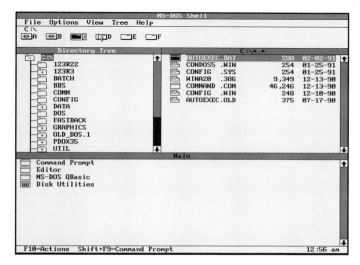

Fig. 2.8

The scroll box for the Directory Tree area.

To scroll down a list, move the mouse pointer to the black area of the scroll bar, below the scroll box. The scroll box moves down, and the list scrolls down to display another part of the list. After you click below the scroll box, the display scrolls down to show the bottom of the list of directories (see fig. 2.9).

To scroll one line up or down, move the mouse pointer to the up scroll arrow or the down scroll arrow, and click. For every click, the display scrolls one line, and the scroll box moves correspondingly.

Figure 2.10 shows the BATCH directory, which contains 129 files. The scroll box indicates that only about one-tenth of the files in the list are visible.

When a list is long, you can scroll swiftly by dragging the scroll box. Move the mouse pointer to the scroll box, hold down the mouse button, and move the mouse up or down. As long as you keep the mouse button pressed as you move the mouse, the list scrolls, and the scroll box moves with the mouse pointer.

You can also scroll with the keyboard. Press [Tab⇄] to select the area you want;
then press [↓] or [↑] to move the selection cursor one item at a time. Press
[PgDn] and [PgUp] to scroll a full screen at a time.

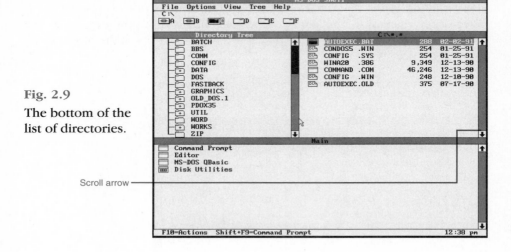

Fig. 2.9
The bottom of the
list of directories.

Scroll arrow ─────

Fig. 2.10
The BATCH
directory.

Getting On-Line Help

The DOS Shell has extensive on-line help available at all times. To get help for a specific menu option, command, or dialog box, choose the item for which you want help, and press F10. For general help with the Shell, commands, procedures, and the keyboard, select the **Help** from the menu bar (see fig. 2.11). The options listed in the **Help** pull-down menu provide additional help.

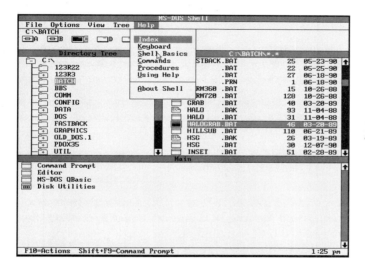

Fig. 2.11
The Help menu.

Figure 2.12 shows additional help topics available for the Shell. To use the mouse to get help on any of the general help, double-click the topic. If you are using the keyboard, press Tab⇆ to underline a topic; then press ↵Enter.

Each Help screen contains the following command buttons:

Command Button	*Description*
Close	Cancels Help
Back	Displays preceding Help screen
Keys	Displays help on keyboard keys
Index	Displays the Help index
Help	Displays the Help screen on using the Help system

2

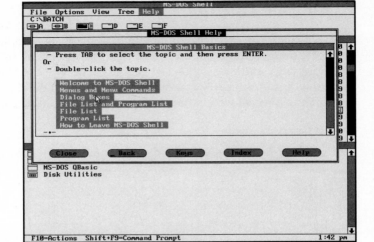

Fig. 2.12
Shell Basics help
topics.

Chapter Summary

This chapter introduces you to the *why* of DOS—why it exists and why you
need to learn it. You have learned the role DOS and ROM play in booting the
computer and have practiced two different methods for starting your computer session. Finally, the chapter discusses the DOS Shell and illustrates
some of the basic skills required to use it.

Testing Your Knowledge

True/False Questions

1. The three components of DOS are the DOS Shell, the command line,
   and the keyboard.

2. A warm boot takes more time to execute than a cold boot.

3. All commands available at the prompt level can also be accessed
   through the DOS Shell.

4. An ellipsis after a menu item indicates that a dialog box will appear
   asking for more information.

5. Scroll boxes are used when all the information cannot fit on one
   screen.

Multiple Choice Questions

1. The operating system handles which of the following during the boot process?
    A. Initial direction of electricity to ROM
    B. Testing internal memory (RAM)
    C. Checking the printer, monitor, and keyboard
    D. A and B
    E. B and C

2. Which of the following files is responsible for interpreting commands entered via the keyboard at the DOS level?
    A. AUTOEXEC.BAT
    B. READ.ME
    C. DIR
    D. COMMAND.COM
    E. MENU.BAT

3. Which of the following keys cancels most menu selections in the DOS Shell?
    A. **space bar**
    B. Del
    C. Home
    D. Esc
    E. ↵Enter

4. Which of the following keys can be used from the keyboard to navigate between items in a dialog box in the DOS Shell?
    A. Tab
    B. ⇧Shift + Tab
    C. an arrow key
    D. none of the above
    E. A, B, and C

5. What is the best way to exit the DOS Shell?
    A. Do a warm boot.
    B. Do a cold boot.
    C. Press the Esc key about 12 times.
    D. Make the appropriate menu selection.
    E. none of the above

Fill-in-the-Blank Questions

1. A warm boot requires _____ time than a cold boot.
2. A(n) _____ after a menu item indicates that a dialog box will appear, prompting for more information.
3. Files with _____EYE_____ and _____OOM_____ extensions are generally used to start a program.
4. The _____SCAP_____ key cancels menu selection in the DOS Shell.
5. The pointing device that is helpful for use with the DOS Shell is called a(n) _____MOUSE_____.

Review: Short Projects

1. Portraying a Floppy Disk

   How does the Shell portray a floppy disk? (Draw a picture.)

2. Portraying a Hard Disk

   How does the Shell portray a hard disk? (Draw a picture.)

3. Portraying the Mouse Shapes

   Create a table, drawing the various mouse shapes and describing their corresponding meanings for the DOS Shell.

Review: Long Projects

1. Printing Copies of the AUTOEXEC.BAT and CONFIG.SYS Files

   Use the Print option on the File menu to print copies of the files AUTOEXEC.BAT and CONFIG.SYS. Next to each line of the printout of each file, make a note, listing what you think that line means.

   Leave the space blank if you have no idea what to put.

   Save the printouts and your notes. In a future project, you will be asked to review your work in this project and make any corrections you think necessary.

2. Drawing a Map of the Current Directory Tree

   Use information presented when you use DOSSHELL to draw a map of your computer's current directory tree. List only the directory names; do not list the individual files. Save the drawing for a future project.

A First Look at DOS Commands

Thursday Homework

Have you ever met someone who intimidated you? Maybe the person was taller, spoke in an authoritative voice, or projected an unusually strong image. Intimidation happens to everyone. Often, though, initial impressions have little to do with the true nature of a person. The same is true of DOS. Despite what you may hear, DOS is designed to be well-mannered. This chapter makes issuing DOS commands from the DOS command line easy to master.

Objectives

1. To Understand the Parts of a DOS Command
2. To Issue Commands from the DOS Prompt
3. To Use the DIR Command
4. To Use Wild Cards Appropriately
5. To Format High- and Low-Density Disks
6. To Use the COPY Command
7. To Use Other Basic DOS Commands

3

Key Terms in This Chapter	
Command	A collection of characters that tell the computer what you want it to do. Most commands are mnemonics of English words, with single numbers or letters often added as optional instructions.
Command line	The line at the DOS prompt on which you type DOS commands rather than use the Shell.
Syntax	The set of rules you follow when issuing commands.
Parameter	Additional information after the command name that refines what you want the DOS command to do.
Switch	A part of the command that turns on an optional instruction or function.
Delimiter	A character that separates the parts of a command. Common delimiters are the space and the slash (/).
Path	A DOS command parameter that tells DOS where to find a file or carry out a command.
Wild card	A character you substitute for another character or characters.
Format	To prepare a disk for data storage.
Volume label	A name that identifies a particular disk.
Track	A circular section of a disk's surface that holds data.
Sector	A section of a track that acts as the disk's smallest storage unit.
Allocation unit	A group of sectors that DOS uses to keep track of files on the disk.
Unformat	To recover the files on a disk after it has been formatted.

Objective 1: To Understand the Parts of a DOS Command

To tell DOS what you want it to do, you enter DOS commands. Commands are letters, numbers, and acronyms separated—or *delimited*—by certain other characters. Stripped of jargon, DOS command usage is like telling your dog to "sit," "heel," and "stay." Moreover, you can tell your computer to "sit and bark" at the same time. DOS commands frequently, although not exclusively, use slash marks (/) to indicate such additional instructions.

A command you give to DOS is similar to a written instruction you may give to a work associate, but with DOS you must be precise. People use interconnecting words and inferences that the human brain can easily grasp. DOS knows only what its developers have programmed it to understand.

When you type at the DOS prompt a command in its proper form, or *syntax*, the DOS command and any additional parameters communicate your intent. Both relay the action you want to perform and the object of that action. Remember that your main tool for communication with your PC is the keyboard. Your PC is ready to work for you, but it doesn't respond to humor, anger, frustration, or imprecise syntax.

Assume that you have an assistant with a limited vocabulary. If you want a sign on a bulletin board duplicated for posting on another bulletin board, you may instruct your assistant to "Copy sign A to sign B. Make sure that the copy is free from errors."

Similarly, if you want DOS to duplicate the data disk in drive A on a disk you have placed in drive B, you give DOS the following instruction:

diskcopy a: b:

To have DOS—the efficient helper—compare the copy and the original, you type the following:

diskcomp a: b:

DISKCOPY and DISKCOMP are good examples of DOS commands that are clearly named to explain the activity they execute. The letters A and B specify the disk drives you want DOS to use.

Although you may rarely use more than a basic set of 10 DOS commands, DOS recognizes and responds to dozens of commands. The most common of these commands are built into the command processor (COMMAND.COM) and are

3

instantly available at the system prompt. Because these commands are always ready for use, they are called *internal* commands. When you execute commands using the Shell menu, you are really executing internal commands.

Other commands are stored as individual programs in the DOS directory of your hard disk. These *external* commands are located, loaded, and executed when you type them at the system prompt and press ⏎Enter. External commands can execute from the system prompt in the same way as internal commands. The commands in the Main and Disk Utilities groups in the Shell are examples of external commands.

Learning the ins and outs of using DOS commands takes practice. DOS commands follow a logical structure that is far more rigid than a casual conversation with a neighbor. The Shell masks some of this rigidity, however.

The strength of DOS is that after you understand its rules, it is very easy to use. Commands conform to standard rules in their command-line structure. When you understand the concepts behind the commands, you can generalize rules to different commands.

To feel comfortable with DOS commands, remember the following points:

- DOS requires you to follow a specific set of rules, or syntax, when you issue commands.
- Parameters, part of a command's syntax, can change the way a command is executed.

You can think of the command name as the action part of a DOS command. In addition to the name, many commands require or provide for further directions. Any such additions are called *parameters*; they tell DOS what or how to apply the action.

Reading many DOS manuals is like reading a French menu when you don't speak French. Just as the waiter stands over you with a casual, smug attitude, these DOS books make little effort to help you navigate through the menu. In fact, most DOS manuals are agony for the new user, and even experienced users may be driven to frustration in fruitless quests for information.

The next few pages, therefore, explain what components make up a DOS command and, just as important, how to read a DOS manual.

Syntax

Syntax is the structure, order, and vocabulary for typing the elements of the DOS command. Using proper syntax when you enter a DOS command is comparable to using proper English when you speak. DOS must clearly understand what you are typing in order to carry out the command.

When you use the Shell, you do not have to know some of the syntax because the information is built into the menus and program items. You need to know the syntax only to complete the Program Properties dialog box. With the command line, however, you must use correct syntax every time you execute a command.

Unfortunately, many DOS manuals use symbolic form to describe command syntax without revealing what each term means. Simply stated, *symbolic form* is the use of a letter or name for illustrative purposes. A file name used to illustrate a command may be called EXAMPLE.COM, for example, even though EXAMPLE.COM exists only in the mind of the writer. When you enter the real command, you must substitute the actual name of the command for the symbolic one.

Another difficulty with using such books is that they frequently list every *command switch* (option) as though multiple switches are a normal part of the command. In fact, many DOS commands cannot be issued to accept every possible option. Using all the options is like ordering a sandwich with white, rye, whole wheat, and cinnamon-raisin bread. The choice is usually either/or, rather than all. A command should contain nothing more than what you want DOS to do.

Symbolic form is used to describe the entire command line. A DIR command in symbolic form looks like the following:

DIR d:path\filename.ext/W/P/A:(attributes)/O:(order)/S/B/L

In contrast, a command you would actually use may look like the following:

DIR C:/P

As you can see, symbolic notation may confuse rather than enlighten you until you understand the concept behind the form.

Switches

A *switch* is a parameter (a single character preceded by a slash) that turns on an optional command function. In the DIR example, /W and /P are switches. Not all DOS commands use switches, however, and switches may have different meanings for different commands.

Many DOS commands can be typed in several forms and still be correct. Although the simple versions of DOS syntax work effectively, most DOS manuals show the complete syntax for a command, which can be confusing. The complete syntax for the DIR command looks like this:

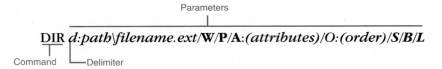

3

Table 3.1 breaks down the sample command piece by piece. Keep in mind that it is the symbolic form of all the alternatives.

Table 3.1 DOS Command Elements	
Element	*Explanation*
DIR	Typing **DIR** at the DOS prompt instructs DOS to run a directory listing. With no switches, the listing appears in vertical form and shows the file names, file size, and the time and date of each file's last modification. When typing the command, do not leave a space between the > symbol and the command.
d:	Represents the drive containing the directory listing. If the system prompt indicates the drive you want, you don't need to specify a drive. If you want a directory listing of another drive, type the appropriate drive letter.
path	Directs DOS to the directory that contains the files you want listed. If you want a directory listing of the files in the current directory, you don't need a path. In fact, the path is rarely used.
filename.ext	Symbolic form meaning "substitute the file name." DOS never allows more than one file with the same name in any one directory. Other than displaying the file size and the amount of space free on the disk, the term has little effect when you use the DIR command.

Element	Explanation
	The following command, for example, adds the file name parameter:
	DIR C:MYFILE.TXT
	In symbolic notation, MYFILE.TXT is shown as *filename.ext*.
/W	DIR /W causes DOS to display a horizontal (or *wide*) directory listing of file names only rather than the usual single vertical listing. The slash (/) acts as the delimiter, telling DOS that a switch follows.
/P	DOS normally scrolls down a directory from beginning to end without stopping. The /P switch requests a pause when the screen is filled. Pressing a key lists another screen of files.
	Tip: As an alternative to using the /P switch, you can press Ctrl + S or Pause to stop a long directory listing from scrolling off the screen. To resume scrolling, press any key.
/A:*(attributes)*	The /A: switch lists only files with certain file attributes. You can list or exclude from the list hidden files (*h*), directories (*d*), system files (*s*), archived files (*a*), and read-only files (*r*). A minus sign before the attribute excludes files with that attribute. The colon (:) after the A is optional but makes the command easier to understand.
/O:*(order)*	The /O switch specifies the sort order of the listing. The order may be alphabetical by name (*n*), alphabetical by extension (*e*), by date and time (*d*), by size (*s*), or directories grouped before files (*g*). A minus sign before the sort order reverses the order. The colon (:) is optional but makes the command easier to understand.

continues

Table 3.1 Continued	
Element	*Explanation*
/S	This switch lists the files in all subdirectories.
/B	This switch lists only the file names with no spaces between the name and the extension.
/L	This switch displays the listing in lowercase instead of uppercase.

DIR /W displays the directory listing in a wide arrangement, but you lose information about individual files (see fig. 3.1).

DIR /P displays the directory listing page by page (see fig. 3.2). The command itself scrolls off the screen when you press a key to continue listing the directory.

DIR /A:D lists only the directories (see fig. 3.3).

When issued at the root directory of a bootable disk, DIR /A:S lists the DOS system files (see fig. 3.4). These files normally do not display on a directory listing.

DIR /A:D/O:N lists only directories and sorts them by the name of the directory (see fig. 3.5).

DIR /W/O:E/S lists the files in wide format, in order by file extension, including files in subdirectories. In figure 3.6, the first directory of this listing has scrolled off the screen.

Fig. 3.1
Displaying a wide
directory listing.

```
C:\BATCH>DIR /W

 Volume in drive C is DRIVE_C
 Volume Serial Number is 1635-9ECA
 Directory of C:\BATCH

[.]            [..]           123.BAT        ADDPATH.BAT    AF.BAT
BLUE.BAT       PDOX.BAT       BNCOMM.BAT     BOTTOM.TXT     CEDB.BAT
CL.BAT         CLEAR.BAT      COMM.BAT       DATA.BAT       FB.BAT
FORM360.BAT    FORM720.BAT    HSG.BAT        HALOGRAB.BAT   INSET.BAT
INSETUP.BAT    MACE.BAT       ORGPATH.BAT    PC.BAT         PM.BAT
SETPATH.BAT    PU.BAT         R22.BAT        R22BUD.BAT     R22D.BAT
R22DB.BAT      R22H38.BAT     R22HOME.BAT    R22PMG.BAT     R22TAX.BAT
R3.BAT         REV.BAT        SQZ.BAT        START.BAT      SUB.BAT
TM.BAT         TM2.BAT        SETPATHW.BAT   SETPATHP.BAT   VP.BAT
WORD.BAT       WIN.BAT        WORKS.BAT      PDOX.PIF       Y
EDS.BAT        TSR.BAT        PX.BAT         PXPATH.BAT     HALO.BAT
STARTCL.BAT    SNAP.BAT       SAVE.BAT       BRIGHT.BAT     NORMAL.BAT
PROMPTN.BAT    R22A.BAT       WORDE.BAT      WDIR.BAT
        64 file(s)        3124 bytes
                     6025216 bytes free

C:\BATCH>
C:\BATCH>
```

```
Volume in drive C is STACVOL_DSK
Directory of C:\

7UP         <DIR>    01-22-93    2:35p
PAAT        <DIR>    02-10-93    4:07p
ALDEO       <DIR>    01-25-93    5:15p
AMIPRO      <DIR>    01-22-93    2:08p
ANYKEY      <DIR>    01-09-93    2:54p
BACKUP      <DIR>    01-09-93    2:54p
BIN         <DIR>    01-22-93    2:35p
COLLAGE     <DIR>    01-22-93    2:08p
CPTOOLS     <DIR>    01-09-93    2:44p
DEV         <DIR>    01-22-93    2:35p
DOS         <DIR>    01-09-93    2:43p
HIJAAK      <DIR>    01-22-93    2:10p
MACH32      <DIR>    01-09-93    2:51p
MACLINK     <DIR>    01-22-93    2:10p
MSMOUSE     <DIR>    01-09-93    2:54p
NET         <DIR>    01-22-93   10:23a
NU          <DIR>    01-22-93    2:09p
OLC         <DIR>    01-09-93    2:51p
PCTOOLS     <DIR>    01-22-93    2:10p
Press any key to continue . . .
```

Fig. 3.2
Pausing the
directory
listing.

3

```
C:\DATA>DIR/A:D

 Volume in drive C is STACVOL_DSK
 Directory of C:\DATA

.            <DIR>    03-23-93    2:34a
..           <DIR>    03-23-93    2:34a
TAXES        <DIR>    03-23-93    2:34a
PERSONAL     <DIR>    03-23-93    2:35a
HOMEFIN      <DIR>    03-23-93    2:35a
WORK         <DIR>    03-23-93    2:35a
NLETTER      <DIR>    03-23-93    2:35a
DOCS         <DIR>    03-23-93    2:35a
        8 file(s)           0 bytes
                    160866304 bytes free

C:\DATA>
```

Fig. 3.3
Listing
directories
only.

```
C:\>DIR /A:S

 Volume in drive C is STACVOL_DSK
 Directory of C:\

IO       SYS     33430 11-11-91    5:00a
MSDOS    SYS     37394 11-11-91    5:00a
MIRORSAV FIL        41 02-11-93    1:42p
        3 file(s)       70865 bytes
                    160858112 bytes free

C:\>
```

Fig. 3.4
Listing the DOS
system files.

```
C:\DATA>DIR /A:D/O:N

 Volume in drive C is STACVOL_DSK
 Directory of C:\DATA

.              <DIR>       03-23-93    2:34a
..             <DIR>       03-23-93    2:34a
DOCS           <DIR>       03-23-93    2:35a
HOMEFIN        <DIR>       03-23-93    2:35a
NLETTER        <DIR>       03-23-93    2:35a
PERSONAL       <DIR>       03-23-93    2:35a
TAXES          <DIR>       03-23-93    2:34a
WORK           <DIR>       03-23-93    2:35a
        8 file(s)             0 bytes
                    160874496 bytes free

C:\DATA>
```

Fig. 3.5
Listing directories
sorted by directory
name.

Among the many thousands of possible DIR command possibilities, you can
type the DIR command in the following ways:

dir

dir /p

dir /w

dir /w/p

dir a:

dir a:/p

dir /p/a:d

dir /w/o:d

dir /b

dir /a:e/s

dir /l

```
HP4-Y.CAR      HP4-Z.CAR      LBP8-S1.CAR    CHKLIST.CPS    FONTGEN.EXE
IFL.EXE        LIB123.EXE     LGOTHIC.IFL    LINEPR.IFL     HELV.IFL
P-AVGARD.IFL   P-BOOKMN.IFL   P-CHNCRY.IFL   P-CNTURY.IFL   P-DINGB.IFL
P-HLV.IFL      P-HLV-C.IFL    P-PALTNO.IFL   P-TIMES.IFL    PRESTIGE.IFL
TMSRMN.IFL     SWISS.IFL      DUTCH.IFL      COUK.IFL       XSYM.IFL
FONTS.LBR      FONT.LST       PRINTER.LST    LS100003.SPD   LS100011.SPD
LS100419.SPD   LS128556.SPD   AA0003.TDF     AI0011.TDF     BK0419.TDF
EM8556.TDF
        56 file(s)     1031963 bytes

Directory of C:\123R23\WYSYGO

[.]            [..]           EATL.CNF       CATL.CNF       HATL.CNF
CHKLIST.CPS    ETOC.IMP       ECHAP1.IMP     ECHAP2.IMP     ECHAP3.IMP
ECHAP4.IMP     ECHAP5.IMP     ECHAP6.IMP     CCHAP1.IMP     CCHAP3.IMP
CCHAP5.IMP     CCHAP6.IMP     HTOC.IMP       HCHAP1.IMP     HCHAP2.IMP
HCHAP3.IMP     HCHAP4.IMP     HCHAP5.IMP     HCHAP6.IMP     CCHAP2.IMP
CCHAP4.IMP     CTOC.IMP       TUTORIAL.OVR
        28 file(s)      966579 bytes

Total files listed:
       201 file(s)    5848820 bytes
                     160041728 bytes free

C:\123R23>
```

Fig. 3.6
Sorting the file
listing by file
extension.

At this point, you may think that the DIR command is much too complicated for anyone except the most advanced DOS experts. Actually, DIR is the easiest command to use because you can ignore most of this information. In the Shell, the Directory Tree and the Files List areas display this information in a slightly different format. Most of these DIR options are available by choosing **Display** from the **Options** menu.

You need not worry about your PC's pulling something sneaky when you type a command. No command you type is executed until you press ⏎Enter. Operating DOS is simpler than you may expect. Just remember that as you gain experience, you can begin to use even potentially dangerous commands in a routine manner.

To show you how routine DOS commands can be potentially dangerous, consider the command that strikes fear in every new PC user: the DEL (or ERASE) command.

3

Experienced computer users make more mistakes with DEL than with any other command because they occasionally zip through their work in a careless manner. With most commands, DOS is forgiving. After all, no harm is done if you accidentally type a request to view a directory of a different drive than you really want. Retyping the instruction properly is simple. The DEL command, however, is another matter.

The DEL and ERASE commands are identical. These commands erase from a designated directory a file or group of files. If you are careless when issuing this command, you can get into big trouble by deleting files unintentionally. DOS 5 and 6 both offer UNDELETE commands to retrieve accidentally deleted files, but using caution before a deletion is much easier than fixing an unintentional deletion.

Objective 2: To Issue Commands from the DOS Prompt

In the Shell, you select a command from a menu or a list of programs in a group. At the command line, you type the command name.

Typing the Command Name

When you type a command, you enter the DOS command directly after the prompt, with no space after the greater-than sign (>). If the command has no parameters or switches, press ⏎Enter after the last letter of the command name. You type the directory command, for example, as **dir** at the prompt and then press ⏎Enter.

Adding Parameters

When you enter parameters that are not switches, this book shows them in two ways: lower- and uppercase. For the lowercase text, you must supply the value. The lowercase letters are shorthand for the full names of the parts of a command. For the uppercase text, you enter letter-for-letter what you see.

Remember that you *delimit*, or separate, parameters from the rest of the command. Most of the time the delimiter is a space, but DOS accepts other delimiters such as the comma (,), the backslash (\), and the colon (:). The examples in this book provide the correct delimiter.

If the command in the example has switches, you can recognize them by the preceding slash (/). Always enter the switch letter as shown.

Ignoring a Command (Esc)

Don't worry if you mistype a command. DOS does not act on a command until you press ⏎Enter. You can correct a mistake by using the arrow keys or ◄Backspace to reposition the cursor. To clear the last entry, press Esc and retype the command.

Just remember that these line-editing and canceling tips work only before you press ⏎Enter. Some commands can be successfully stopped with the Ctrl + C or Ctrl + Break sequence, but checking that the command is typed correctly is always good practice.

Executing a Command

Because ⏎Enter is the action key for DOS commands, you should make it a habit to pause and read what you have typed before you press ⏎Enter.

During the processing of the command, DOS does not display any keystrokes you may type. It does remember your keystrokes, however, so be aware that any characters you type may end up in your next command.

Using DOS Editing Keys

When you type a command line and press ⏎Enter, DOS copies the line into an input *buffer*, a storage area for commands. With F3, you can pull the last command line from the buffer and use the line again. This feature is helpful when you want to issue a command that is similar to the last command you used. Table 3.2 lists the keys you use to edit the input buffer.

Table 3.2 DOS Command-Line Editing Keys	
Key	*Action*
Tab⇥	Moves cursor to the next tab stop
Esc	Cancels the current line without changing the buffer
Ins	Enables you to insert characters in the line

continues

Key	Action
	Table 3.2 Continued
Del	Deletes a character from the line
F1 or →	Copies one character from the preceding command line
F2	Copies all characters from the preceding command line up to the next character you type
F3	Copies all remaining characters from the preceding command line
F4	Deletes all characters from the preceding command line up to, but not including, the next character typed (opposite of F2)
F5	Moves the current line into the buffer without executing the line
F6	Produces an end-of-file marker when you copy from the console to a disk file

Objective 3: To Use the DIR Command

As figure 3.7 illustrates, the DIR command displays much more than a list of file names. As your computing expertise grows, you will find many uses for the information provided by the full directory listing.

Defining the DIR Command

A directory is a list of files. With the DIR command, you get the following information:

- Drive letter
- Volume label
- Volume serial number assigned by DOS
- Name of the directory

- Five columns of information about the files (refer to fig. 3.7)
- Number of files listed
- Total number of bytes in the files listed
- Amount of unused space on the disk

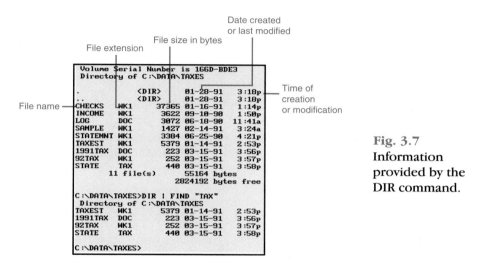

Fig. 3.7
Information
provided by the
DIR command.

Try the DIR command now. Type **dir**, and press ⏎Enter.

You just told DOS to list the files on the logged drive. You can also type **dir a:** to specify drive A or **dir c:** to list the files on drive C. (A and C are drive parameters.) If you don't specify a drive, DOS uses the current drive.

In the Shell, you change the current drive by selecting a drive from the list of disk drive icons. At the command line, you change the current drive by typing a drive letter and colon and pressing ⏎Enter. By typing **a:** at the DOS prompt, for example, you change the current drive to A. A disk must be in a drive before DOS can make it the current drive. Changing the current drive enables you to switch between a hard disk and a floppy disk.

Exercise 3.1: Issuing the DIR Command

In this exercise, you practice viewing the files on your computer's drive C. Follow these steps:

1. Boot your computer, if necessary.
2. To view the files on drive C one screen at a time, type `dir c: /p` at the DOS prompt, and press `↵Enter`.
3. As the file names scroll to a stop, press `↵Enter` to see each new screen until all file names have been displayed.
4. Now view the files on drive C in the five-column format. Use the command listing in table 3.1 to determine the appropriate switch for doing this.
5. If you have a formatted disk in drive A, issue the command to view a list of files on drive A, again using the command listing in table 3.1 for help.

Controlling Scrolling

Scrolling refers to the way a screen fills with information. As the screen fills, the lines of the display scroll off the top of the screen. To stop a scrolling screen, press `Ctrl`+`S`; then press any key to restart the scrolling. On enhanced keyboards, press `Pause` to stop the scrolling.

Assigning File Names and Extensions

The file name contains two parts: the name and the extension, with a period separating the file name from its extension. In the directory listing, however, spaces separate the file names and extensions.

In any single directory, each file must have a unique name. DOS treats the file name and the extension as two separate parts. The file names MYFILE.WK1 and MYFILE.ABC are unique because each file has a different extension. The file names MYFILE.WK1 and YOURFILE.WK1 also are unique. Many DOS commands make use of the two parts of the file name separately. For this reason, giving each file an extension is a good idea.

File names can help you identify the contents of a file. DOS file names can contain only eight alphanumeric characters, plus a three-character extension. With this built-in limit, meeting the demand of uniqueness and meaningfulness needed for some file names can require ingenuity.

DOS also is specific about which characters you may use in a file name or an extension. To be safe, use only letters of the alphabet and numbers, not spaces or a period. DOS truncates excess characters in a file name.

Understanding File Size and Date/Time Stamps

In the directory listing, the third column shows the size of the file in bytes. This measurement is an approximation of the size of your file, which may actually contain a somewhat lower number of bytes than shown. Because computers reserve blocks of data storage for files, files with slightly different data amounts may have identical file-size listings. This disk-space allocation method also explains why a memo with only five words may occupy 2K of file space.

The last two columns in the directory listing display a date and a time. These entries represent the time you created the file or, with an established file, the last time you altered the file. Your computer's internal date and time are the basis for the date and time stamp in the directory. As you create large numbers of files, the date and time stamps become invaluable tools in determining the most recent version of a file.

Objective 4: To Use Wild Cards Appropriately

Technically, the *wild card* in a file specification is a character that represents one or more characters. In DOS, the question mark (?) represents any single character. The asterisk (*) represents one or more characters in a file name.

Although the command, COPY A:*.BAT C:/V, may seem unintelligible, it simply states the following:

- This is a copy function.
- From the disk in drive A, copy all files (designated by the asterisk) that end with the extension BAT.
- Place a copy of these files in drive C.
- Verify (represented by /V) that the copy is the same as the original.

Using Wild Cards in the DIR Command

The following text provides examples of the use of wild-card characters with the DIR command.

One form of the DIR command looks like the following:

DIR *d:filename.ext*

When you use DIR alone, DOS lists all files in the current directory. When you use DIR with a file name and extension parameter, DOS lists files that match the parameter. In place of *filename.ext*, execute the command as follows:

dir myfile.wk1

The DIR command you just typed tells DOS to list all files in the current directory matching MYFILE.WK1. The directory lists only one file, MYFILE.WK1.

If you want a listing of all files with the WK1 extension, type the following command:

dir *.wk1

DOS lists any file in the current directory with the WK1 extension. File names like MYFILE.WK1 and YOURFILE.WK1 display.

If you issue the command,

dir myfile.*

you may get a listing that includes files such as MYFILE.WK1 and MYFILE.XYZ.

You can use file name extensions to specify the type of file. Correspondence files may have a LET extension, and memo files the extension MEM. This practice enables you to use the DIR command with a wild card to get separate listings of the two types of files.

The ? wild card differs from the * wild card. With the ?, any single character in the same position as the ? is a match. If you issue the command,

dir myfile?.wk1

files such as MYFILE1.WK1 and MYFILE2.WK1 appear, but MYFILE.WK1 doesn't. The same rules apply to other commands that accept wild cards.

Exercise 4.1: Listing Multiple Files with a Wild Card

In this exercise, you practice using wild cards in combination with the DIR command.

1. Boot your computer, if necessary.
2. To view all files ending with the EXE extension, type **dir *.exe**, and press ↵Enter.
3. To view all files that begin with the letter C, type **dir c*.***, and press ↵Enter.
4. To view all files for which the last two letters of the extension are *OM*, type **dir *.?OM**, and press ↵Enter.

Objective 5: To Format High- and Low-Density Disks

Now that you have had some practice entering DOS commands, this section shows you how to format a floppy disk so that you can begin storing files for your own use.

Understanding Floppy Disks

A floppy disk is a Mylar pancake in a plastic dust cover. The Mylar disk is covered with magnetic material similar to the metallic coating on recording tape. A disk must be formatted before you can use it. Although you can buy formatted disks, few stores currently carry them.

The DOS FORMAT command performs the preparation process for disks. You simply enter the command, and FORMAT checks for disk defects, generates a root directory, sets up a storage table (called a *file allocation table*), and alters other parts of the disk.

The unformatted magnetic disk is like unlined paper—hardly a good medium for proper magnetic penmanship. Although blank pages are fine for writing casual notes, you may wind up with wandering, uneven script; most writers need lines to serve as guides. Although *you* can write on unlined paper, DOS is not so flexible. Your computer cannot use a disk at all until it is formatted.

Formatted disks can be compared to lined paper, with horizontal lines subdivided by vertical lines (see fig. 3.8). As lines on paper are guides for the writer, tracks and sectors on a disk are guides for the computer. Because the storage medium of a spinning disk is circular, the "premarked lines," or magnetic divisions are called *tracks* and are placed in concentric circles. These tracks are further subdivided into areas called *sectors* (see fig. 3.9).

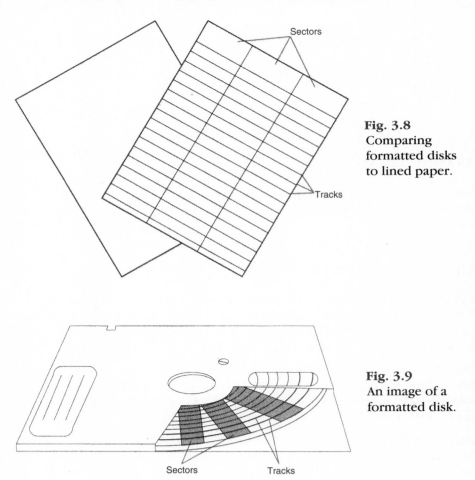

Fig. 3.8
Comparing formatted disks to lined paper.

Fig. 3.9
An image of a formatted disk.

When you format a blank disk, DOS magnetically *encodes*, or marks, tracks and sectors onto the disk surface, positioning them according to the type of drive you have. DOS then stores data in these sectors and uses the track and sector numbers to find and retrieve information.

Understanding Different Floppy Disks

In order to buy the appropriate disks for your computer, you need to understand the types of floppy disks that are available.

Unfortunately, the various types of disks can be described in many different ways. If you walk into a computer store to buy 1.44 megabyte disks, you may find that the boxes of disks tell you everything but the capacity. With the following information, however, you can understand how to buy the type of disks you need.

Different types and sizes of floppy disks have different numbers of tracks, sectors, and capacities. Over the years, the maximum capacity of floppy disks has increased steadily. In 1981, a floppy disk could hold 160K (kilobytes). Today, the maximum capacity is usually 1.44M (megabytes), but 2.88M floppy disks are starting to appear.

Two sizes of floppy disks are available: 5 1/4 inch and 3 1/2 inch. The larger disks were available first and are almost a rarity these days. The smaller disks have a higher capacity, have rigid cases that protect the disk from damage, and are more reliable. The size of the disks is easy to determine from the size of the box, and the size is clearly marked on the box. This part is easy.

Floppy disks can be *single-sided* or *double-sided*. Double-sided disks have tracks on both sides of the disk and therefore have twice the capacity of single-sided disks. Although only double-sided disks are used in today's PCs, some computer stores may still sell single-sided disks for very old PCs and other computers. Boxes of floppy disks should be labeled *two-sided*, *2S*, *double-sided*, or *DS*. So far so good.

Floppy disks also come in different *densities*, a measure of how closely the bytes of information are placed on the disk. Today, floppy disks are *double density* or *high density*. (High density is sometimes also called *quad density*.) Single-density disks were used only in older types of computers, not in PCs. Double-density disks usually are labeled *DD* or *2DD* (the *2* indicates double-sided). High-density disks usually are labeled *HD* or *2HD*.

Sectors on all types of disks hold 512 bytes (.5K). Disks have different capacities because they may have more sectors per track or more tracks per side.

The following list summarizes the standard floppy disks that are readily available:

Disk Type	Sectors per Track	Tracks per Side	Capacity
5 1/4-inch			
Double density	9	40	360 K
High density	15	80	1.2 M
3 1/2-inch			
Double density	9	80	720 K
High density	18	80	1.44 M

720K disks are often labeled *1M*, and 1.44M disks are often labeled *2M*. These numbers refer to the unformatted capacity of the disk and add to the confusion.

Should you remember all this information about floppy disks? No, the computer and DOS take care of most of it for you. You do need to know, however, what type of disk is required by your computer.

Matching Floppy Disks and the Disk Drive

Just as there are different types of floppy disks, there are different types of disk drives. The most obvious variation is in disk drive size, which can be 5 1/4 inch or 3 1/2 inch. If you have 5 1/4-inch drives, you cannot use 3 1/2-inch floppy disks, and vice versa. If you are lucky enough to have one drive of each size, you can use both disk sizes.

Size is not the only consideration, however. Disk drives also have a maximum capacity. A 5 1/4-inch drive may be a standard (360K) drive or a high-capacity (1.2M) drive. A 1.2M drive can read and write 360K disks, but a 360K drive cannot read or write 1.2M disks. The 1.2M floppy disks fit in the drive, but you cannot use them.

Although you can use 360K floppy disks in 1.2M drives, the results can be unreliable. If you write to a 360K disk in a 1.2M drive, you should have no trouble reading that disk in the same drive. You may have trouble reading the disk in a 360K drive, however. This problem is intermittent and occurs more often with older drives. If you run into this problem when copying files to a floppy disk to be read by another computer, try using another floppy disk or format the disk again.

3

The same situation exists with 3 1/2-inch drives. If you have a 720K drive, you cannot use 1.44M disks. If you have 1.44M drives, you can use both types of 3 1/2-inch floppy disks.

Formatting Floppy Disks

As formatting your disks becomes a routine task, remember to use care. Formatting clears all information that a disk contains. If you format a used disk, all the information stored on that disk disappears, so you must be careful not to format disks with files you want to keep. Labeling your disks can help you avoid such a mishap. Another precaution is to *write-protect* disks that hold important information by adding tape tabs on 5 1/4-inch disks or setting the write-protect switch on 3 1/2-inch disks. Write-protecting a disk is like pushing out the tabs on a cassette tape. When the write-protect tab is on, the write head of the disk drive cannot position itself to write.

In any case, always check the list of files before formatting a used floppy. Checking the files on a disk is like reviewing what you have written on a blackboard before erasing it.

Before you begin, you need to have a floppy disk ready to format. To avoid mistaking formatted disks for unformatted disks, place some indicator on each disk you format. The indicator may be as simple as a dot, a check mark, or the letter *F* for *formatted*. When you buy floppy disks, adhesive labels are included. A simple way to keep track of formatted disks is to put a label on each disk that you format. Then you know that a disk without a label has never been formatted.

If you have two floppy drives, you can format disks in drive A or B. If you have only a single floppy drive, you format the disk as drive A.

Exercise 5.1: Formatting a Disk

In this exercise, you practice formatting a 3 1/2-inch floppy disk.

1. Boot your computer, if necessary.
2. Obtain a new, unformatted disk and insert it into drive A. The metal end goes in first, with the brand name of the disk up. You hear a clicking sound when the disk settles into the drive.
3. At the DOS prompt, type **format a:**, and press ⏎Enter.

4. DOS prompts you to insert a new disk into the drive and press any key
   when ready. You have already put the new disk into the drive, so just
   press ⏎Enter to begin formatting.

   FORMAT checks the existing disk format. If the disk has never been
   formatted, the program proceeds with the format. As the disk is
   formatted, the percentage of the disk that has been formatted is
   displayed. When the process is complete, you see the following
   message:

   ```
   Format complete
   Volume label (11 characters, Enter for none)?
   ```

   DOS reserves a few bytes of space so that you can place an electronic
   identification, called a *volume label*, on each disk.

5. Volume labels can be up to 11 characters in length (as the message
   instructs) and can contain letters A–Z, numerals 0–9, and the ~ ! @ #
   $ ^ & ( ) _ { } 'special characters. Volume labels are optional but quite
   useful if you forget to put a sticky label on the outside of the disk.

   Type **mydisk**, and press ⏎Enter.

6. After you type the volume label, DOS displays detailed information
   about the formatted disk and then asks whether you want to format
   another disk. If you do not want to format any more disks, press N
   and then ⏎Enter. Press Y and then ⏎Enter if you want to format
   another disk while FORMAT is still loaded into the computer's
   memory.

   To continue with this exercise, press N; then press ⏎Enter.

7. On the gummed paper label that came with your disk, write the name
   MYDISK. Eject your formatted disk from the A drive, and paste the
   label to the disk, as shown on the side of the disk box.

Understanding Safe Formatting

When you format a disk that has been formatted previously, FORMAT saves
some information that enables you to *unformat* the disk later. Remember,
when you format a disk, you tell DOS to clear the disk completely so that you
can use it again. If you format a floppy disk by mistake, however, and do not
put any other files on the disk, you can still unformat the disk to recover these
files. This process is called *safe formatting*.

Formatting Different Types of Disks

At times, DOS may not know what type of disk you want to format. If you have a 5 1/4-inch 360K drive or a 3 1/2-inch 720K drive, you have no problem because these drives can use only one type of disk.

High-capacity disk drives can format more than one type of disk, however. If you have a 1.2M drive, you can use 360K or 1.2M floppy disks. If you have a 1.44M drive, you can use 720K or 1.44M floppy disks. In some cases, you must specify the disk type in the Format dialog box. The default type is always the highest capacity disk that can work in the drive.

When you format a new disk in a 1.44M drive, DOS assumes that you have a 1.44M disk unless you tell DOS otherwise. If you want to format a 1.44M disk in a 1.44M drive, you do not have to do anything special. If, however, you want to format a 720K disk in a 1.44M drive, you must add parameters to the format command. These parameters differ according to the brand of computer you have.

If you have an IBM PS/2 computer and want to format a disk for 720K, use the following command:

FORMAT A: /N:9/T:80

If you have a different type of computer, ask your instructor to assist you in determining which type of disk you have. Use the space provided here to write the format command and necessary switches for your computer.

Looking at the FORMAT Command's Output

After the format is complete and you have entered a volume label, DOS
provides a disk status report that shows the total disk space and total bytes
available. If FORMAT detects bad sectors (bad spots) on the disk, DOS marks
them as unusable. FORMAT also reports how many bytes are unavailable
because of bad sectors. Other information includes the number of bytes each
allocation unit contains, the number of allocation units available on the disk
for storage, and the volume serial number that DOS automatically assigns to
every disk.

The numbers for various sizes of disks vary, as shown in figures 3.10 and 3.11.

Fig. 3.10
Disk in-
formation
for a 720K
disk.

```
Insert new diskette for drive B:
and press ENTER when ready...

Checking existing disk format
Formatting 720K
Format complete

Volume label (11 characters, ENTER for none)? SAMPLE 720K

    730112 bytes total disk space
    730112 bytes available on disk

      1024 bytes in each allocation unit
       713 allocation units available on disk

Volume Serial Number is 0E0A-15D9

Format another (Y/N)?
```

Fig. 3.11
Disk in-
formation
for a 1.44M
disk.

```
Insert new diskette for drive B:
and press ENTER when ready...

Checking existing disk format
Saving UNFORMAT information
Verifying 1.44M
Format complete

Volume label (11 characters, ENTER for none)? SAMPLE 144M

   1457664 bytes total disk space
   1457664 bytes available on disk

       512 bytes in each allocation unit
      2847 allocation units available on disk

Volume Serial Number is 3358-15DC

Format another (Y/N)?
```

Understanding FORMAT's Error Messages

The most common DOS FORMAT error messages are rarely catastrophes. In most cases, they are little more than statements suggesting that you did something wrong or that FORMAT had trouble carrying out the command.

With floppy disks, errors that occur during formatting activity are usually not serious.

If you respond to the Press any key when ready prompt without placing a disk in the disk drive or if the door is open, DOS displays the message:

```
Not ready

Format another (Y/N)?
```

This message simply means that DOS cannot read from the disk drive. Just insert the disk, close the door if necessary, and press Y and then ↵Enter for the format to start. If you specified the wrong disk drive, press N and then ↵Enter. Then execute the command again with the correct drive letter.

Another kind of error involves a write-protection problem. If the disk is write-protected, you get the same message as when you fail to insert a disk in the drive. If you get this error message when a disk is in the drive, take the disk out and check the write-protect tab.

To format a write-protected disk, slide the tab on a 3 1/2-inch disk, reinsert the disk, and press Y at the prompt.

If the FORMAT command detects unusable areas on the disk, you see a line describing the problem in the report. Although not a true error message, a bad-sectors report points out a possible problem with the disk.

The Bytes in bad sectors message means that MS-DOS found on the disk bad sectors that cannot store information. The total amount of free space on the disk is reduced by the number of bytes in the bad sectors.

If you get this message, try reformatting the disk. If it still has bad sectors and is a new disk, you can have your dealer replace the disk, or you can use the disk as is. Before you do either, though, try formatting the disk again.

If you get a very large number of bad sectors, you may have tried to format a 720K disk as a 1.44M disk. DOS tries to format the disk, but after the computer grinds and chugs, you end up with mostly bad sectors. Format the disk again with the /N:9/T:80 switch.

If you reformat a disk that is completely full, no room remains to save the information so that the disk can be unformatted. You see a warning message that the disk cannot be unformatted.

The worst disk-error message that you can get is

```
Invalid media or Track 0 bad - disk unusable
```

This disk may have a scratched surface; DOS was not able to read disk-level information on the first track. You may receive this error message for two other possible reasons. If you try to format a 720K disk as a 1.44M disk, you will get this message. Format the disk again with the appropriate switches.

This message also can mean that the areas on the disk that hold important DOS system data are bad. If you get the `Disk unusable` error message on a new disk, take it back to your dealer. If the disk is old, throw it away. Disks are inexpensive and, in this case, should be discarded. Trying to use a bad disk is being penny wise and pound foolish.

Formatting a Hard Disk—NOT!!

Hard disks are a desirable part of a computer system because of their speed and storage capacity. Just like floppy disks, hard disks must be formatted before you use them. Unless you are familiar with the procedure, however, *do not attempt to format your hard disk!*

Many computer dealers install the operating system on a computer's hard disk before you receive it. If your dealer has installed an applications program, such as a word processing program, do not format the hard disk. If you reformat your hard disk, you will erase all programs and data.

If you ever attempt to reformat your hard disk, first perform a complete backup. Make sure that you are familiar with the RESTORE command (see Chapter 5). You also should have ready a bootable floppy disk that contains a copy of RESTORE. If you must format your hard disk, consult your computer's manual.

Remember that FORMAT erases all the data a disk contains. Always check the directory of the disk you want to format because it may hold data you need. Make a mental note to check the command line thoroughly when you use the FORMAT command.

Objective 6: To Use the COPY Command

COPY enables you to make a duplicate copy of a file on to another disk or to make a duplicate copy of a file, giving the duplicate file a new name.

The standard syntax for the COPY command is as follows:

>COPY *sd:\spath\sfilename.sext dd:\dpath\dfilename.dext /switches*

An easier way to look at the COPY command is with this simplified syntax:

>COPY *source destination*

The *source* is the location and file name of the file to be copied. The *destination* is the desired location and file name of the duplicate file.

By using wild cards with COPY, you can quickly copy multiple files at one time. The best way to learn how to use the COPY command is to look at a few examples.

To copy MYFILE.WK1 from the current directory to drive A, type

>**copy myfile.wk1 a:**

Because no destination file name is specified, the file name is not changed, and MYFILE.WK1 now exists in two places.

To make a duplicate of MYFILE.WK1 in the current directory with the name BACKUP.WK1, type

>**copy myfile.wk1 backup.wk1**

To copy MYFILE.WK1 from the current directory to drive A and change its name to BACKUP.WK1, type

>**copy myfile.wk1 a:backup.wk1**

To copy all files with the WK1 extension from the current directory to drive A, type

>**copy *.wk1 a:**

To copy all files with a file name that starts with BUDGET from the current directory to drive A, type

>**copy budget*.* a:**

The preceding command copies files with names such as BUDGET.WK1, BUDGET1.WK1, BUDGET91.WK1, BUDGET.DOC, and BUDGET91.ZIP.

To copy all files from the current directory to drive A, type

>**copy *.* a:**

Exercise 6.1: Copying AUTOEXEC.BAT to Your Floppy Disk

In this exercise, you practice copying files from the hard disk to the floppy disk you formatted in the preceding exercise. This exercise assumes that the default drive is C.

3

1. Boot your computer, if necessary.

2. Locate the floppy disk labelled MYDISK, and insert it into drive A.

3. To copy the AUTOEXEC.BAT file from drive C to drive A, type **copy c:\autoexec.bat a:** at the C: \>, and press ⏎Enter.

4. DOS informs you that one file has been copied. To verify, type **dir a:** to view the directory of the disk in drive A.

5. When copying files, DOS prevents you from copying a file on to itself. In other words, you must give a new location or a new file name for the destination file. To see what happens when you forget to specify a new location or file name, type **copy c:\autoexec.bat**, and press ⏎Enter.

   DOS responds with the error message Cannot copy file onto itself. In this case, the command does not contain a destination. Remember that error messages are just that—messages that tell you that you have goofed. Read them carefully; they usually tell you exactly what you have done wrong.

   You often need to have a safety copy, called a *backup* copy, of a file to ensure that you don't accidentally erase the file.

6. To copy AUTOEXEC.BAT to a backup copy, type **copy c:\autoexec.bat a:\autoexec.bak**, and press ⏎Enter.

7. Wild cards are very useful when combined with the COPY command. To copy all files ending in BAT from drive C to your disk in drive A, type **copy c:*.bat a:**, and press ⏎Enter.

   DOS copies at least one file, AUTOEXEC.BAT, to drive A. Your hard drive may have several other BAT files, depending on how it is set up.

Objective 7: To Use Other Basic DOS Commands

Most DOS users find it necessary to memorize about ten DOS commands. This chapter has already discussed three of these commands; this section covers three more. The remaining four commands are discussed in Chapter 7.

3

Using the DELETE Command

DEL enables you to remove unwanted or unnecessary files from a disk. This command is very dangerous because it enables you to destroy data, so take extra care to make certain that you enter the command correctly. DOS does not ask you whether you are sure that you want to delete a file as some applications programs do.

The syntax of the DELETE command is as follows:

   DEL *d:path\filename.ext/p*

Exercise 7.1: Deleting AUTOEXEC.BAK from MYDISK

In this exercise, you practice deleting a file from your floppy disk.

1. Insert the disk labeled MYDISK into the A drive.
2. Type **dir a:** to review the contents of the disk. You should have a file called AUTOEXEC.BAK.
3. To delete this file, a duplicate of the AUTOEXEC.BAT file on the disk in drive A, type **del a:\autoexec.bak** at the DOS prompt.
4. Check to make sure that you entered the command correctly—especially the drive—and press ⏎Enter.
5. Type **dir a:** to review the contents of drive A again. AUTOEXEC.BAK should be missing.

Using the RENAME Command

RENAME (REN) enables you to rename files without making a duplicate copy.

The syntax of the RENAME command is as follows:

   RENAME *d:path\oldfilename.ext newfilename.ext*

Exercise 7.2: Renaming a File

In this exercise, you use the RENAME command to change the name of a file.

1. Insert the disk labeled MYDISK into the A drive.
2. Type **dir a:** to review the contents of the disk. You should have a file called AUTOEXEC.BAT.

3. To rename this file to AUTOEXEC.BAK, the file you deleted in the last exercise, type **rename a:\autoexec.bat autoexec.bak** at the DOS prompt.

4. Check to make sure that you entered the command correctly—especially the drive—and press ⏎Enter.

5. Type **dir a:** to review the contents of drive A. AUTOEXEC.BAT should be replaced by AUTOEXEC.BAK.

Using TYPE To View the Contents of a Text File

TYPE is one of the most useful DOS commands you will encounter. It enables you to view the contents of a text file without entering a word processing or text editing package. TYPE works only with *text files*—files that contain no special characters or codes. Most batch files (BAT) are text files; recently, the extension TXT also has come to indicate a text file.

The syntax of the TYPE command is as follows:

TYPE *d:path\filename.ext*

Exercise 7.3: Using TYPE To View AUTOEXEC.BAK

In this exercise, you display the contents of a text file to the screen.

1. Insert the disk labeled MYDISK into drive A.

2. Type **dir a:** to review the contents of the disk. You should have a file called AUTOEXEC.BAK.

3. To view the contents of this file, type **type a:\autoexec.bak** at the DOS prompt. Check to make sure that you entered the command correctly—especially the drive—and press ⏎Enter.

4. DOS lists the contents of AUTOEXEC.BAK on-screen. Remember from Chapter 2 that the AUTOEXEC.* file contains settings for the current computer session. This file is discussed in much greater detail in Chapter 7.

Chapter Summary

In this chapter, you have actually begun to use DOS. You have learned how to format disks, copy files from one disk to another, view files on disk, delete files, rename files, and use wild cards to decrease the number of commands you have to enter. You are now familiar with the basic toolbox commands that every DOS user needs to know. The next chapter presents file organization techniques.

Testing Your Knowledge

True/False Questions

1. In order for the DIR command to work, you must list a drive in the command.

2. The ? wild card is used as the holding place for only one character in a file name.

3. A switch in a command enables you to customize your commands.

4. The number of tracks and sectors on a disk depends on the density of the disk.

5. The TYPE command is useful for displaying the contents of program files.

Multiple Choice Questions

1. Which of the following commands lists the files on the A drive?
   A. DIR A:
   B. A
   C. A:
   D. DIR
   E. none of the above

2. Which of the following sets of commands most efficiently copies the file TEST.DOS from the C drive to a file named TEST.OS on the A drive?
   A. COPY TEST.DOS TEST.OS
   B. COPY TEST.DOS TEST.DOS
      RENAME TEST.DOS TEST.OS
   C. COPY C:\TEST.DOS A:\TEST.OS
   D. COPY C:\TEST.DOS A:\TEST.DOS
      RENAME A:\TEST.DOS TEST.OS
   E. none of the above

3

3. Which of the following files is *not* deleted by the command DEL A:CHPT?.DOC?

    A. CHPT10.DOC

    B. CHPT1.DOC

    C. CHPT2.DOC

    D. CHPT3.DOC

    E. none of the above (all files are deleted)

4. Which of the following files is listed with the command DIR A:CHP*.*?

    A. CH1.DOC

    B. CHP2.DOC

    C. CHP3.DOC

    D. A and B

    E. B and C

5. Of the commands discussed in this chapter, which is most appropriate for viewing the contents of the file AUTOEXEC.BAT?

    A. DIR

    B. COPY

    C. RENAME

    D. TYPE

    E. none of the above

Fill-in-the-Blank Questions

1. The words DIR, COPY, RENAME, and TYPE are all _____ in the DOS command syntax.

2. When you purchase a new box of disks, you must _____ them first before using them.

3. The FORMAT command places _____ and sectors as organizing devices on the disk.

4. The command used to display the contents of a text file is the ____DIR____ command.

5. The switch that causes the output of a DIR command to be displayed in five-column format is ___/W___.

Trabalho next Thursday

Review: Short Projects

1. Formatting a Floppy Disk

   Format a new floppy disk with the volume label option, and name the disk CHAPTER3. Label and save the disk for a future project.

2. Copying Files from the Hard Drive

   Copy all files that end with BAT from the hard drive to this disk. How many files are on your new disk?

3. Using Switches To List Files

   List the most efficient form of DIR and its switches to locate each of the following groups of files (assuming the subject files are in a common directory) *without* displaying other files in the directory:

   A. Locate the files WP.EXE, ONL.EXE, and BIBLIO.EXE. The directory also contains WP.CNF, ONL.DAT, and BIBLIO.FNT.

   B. Locate the files CH01.TXT, CH02.TXT, and CH07.TXT. The directory also contains CH11.TXT and ESSAY.TXT.

   C. Locate the files ACCOUNT1.QFL and ACCOUNT2.QSL. The directory also contains ACCOUNT.QFK and ACCOUNT.QSR.

Review: Long Projects

1. Reviewing the AUTOEXEC.BAT and CONFIG.SYS Printout

   Examine the AUTOEXEC.BAT and CONFIG.SYS files you printed in Chapter 2. You should now be able to add more comments to your notes.

2. Developing a Computer System and Directories

   A friend of yours is running a home secretarial business in which she has been typing reports for 12 different clients. She now wants to automate her business by purchasing a computer. She wants you to recommend what equipment she needs and what software to buy, and she wants you to help her set up the directories on her computer. List the equipment that you recommend and the directories that you will use.

   Hint: She needs a word processor and probably should have a separate subdirectory for each of her clients. She also is worried about how to handle her billing and keep copies of the accounts.

Understanding and Using Directories

4

In this chapter, you explore the concepts of file storage and organization by creating and using directories. This chapter on file "housekeeping" provides you with the skills you need to be an efficient computer user.

Objectives

1. To Understand the Concept of Directories
2. To Navigate within the Directory Structures
3. To Understand DOS Paths
4. To Understand and Explore Subdirectories
5. To Manage Directories on Your Hard Drive

4

Key Terms in This Chapter	
Hierarchical directory	An organizational structure used by DOS to segregate files into different levels.
Tree structure	A term applied to hierarchical directories to describe the concept in which directories "belong" to higher directories and "own" lower directories. Viewed graphically, the ownership relationships resemble an inverted tree.
Directory	An area of the DOS file system that holds information about files and directories. The root directory is the highest directory of DOS's tree structure. All DOS disks have a root directory that DOS (Versions 2.0 and later) creates automatically.
Subdirectory	A subordinate directory created within another directory, also called a directory.
Directory specifier	A DOS command parameter that tells DOS where to find a file or where to carry out a command.
Path name	Another name for the directory specifier. The path name gives DOS the necessary directions to trace the directory tree to the directory that contains the desired commands or files.
Backslash (\)	The character that DOS expects to see in a command to separate directory names. Used alone as a parameter, the backslash signifies the root directory.

Objective 1: To Understand the Concept of Directories

A continuing problem for PC users is remembering where files are stored and how to get to them. The computer's capability of storing millions of files in one place requires some kind of organization. The DOS programmers who worked to develop the technology we now use in hard drives also developed the notion of a directory structure based on the real-life structure of a filing cabinet.

You can think of a hard drive as a file cabinet, a single unit that can store thousands of documents. Each drawer in the file cabinet holds a particular set of documents. In the same way, each directory on a hard drive is set up to store specific files, which correspond directly with the file folders in the file cabinet. If you can work with this mental image, dealing with directories will be much easier for you.

Chapter 3 illustrates file lists of the contents of a disk directory. A directory is more than a file list displayed on-screen, however. A directory is also part of an internal software listing that DOS stores in a magnetic index on the disk. A poorly structured disk directory can turn a hard drive into a bewildering tangle of misplaced files.

4

This chapter explains DOS's hierarchical directory structure and shows you how to use DOS commands to group and organize your files in a logical directory structure.

Objective 2: To Navigate within the Directory Structures

DOS uses directories to organize files on disk. A directory listing contains file information, including the name, size, and date of creation or revision for each file. Computer operators use the directory of a disk to find specific files. DOS also uses the directory information to respond to requests for data stored in the files on disk.

All MS-DOS-based disks have at least one directory, which is usually adequate for the relatively limited capacities of floppy disks. Hard disks, on the other hand, have very large storage capacities, often containing hundreds or even thousands of files. Without some form of organization, you can waste a great deal of time sorting through your disk's directories to find a specific file.

Although floppy disks can use DOS's multiple directory structure, this feature is more important for maintaining order on hard disks. With a bit of foresight, you can store your files in logically grouped directories so that you (and DOS) can locate your files more easily.

MS-DOS Versions 2.0 and higher incorporate the *hierarchical directory* system, a multilevel file structure that enables you to create a filing system. Hierarchical directories are like the library system of storing books in categories broken down by ever-narrower subjects.

Computer professionals use the term *tree structure* to describe the organization of files into hierarchical levels of directories. Try picturing the tree
structure as an inverted tree (see fig. 4.1). You can visualize the file system
with the first-level directory as the root or trunk of the tree. The trunk
branches into major limbs to the next level of directories under the root.
These directories branch into other directories. Directories have files, like
leaves, attached to them. The terms *directory* and *subdirectory* are inter-
changeable.

4

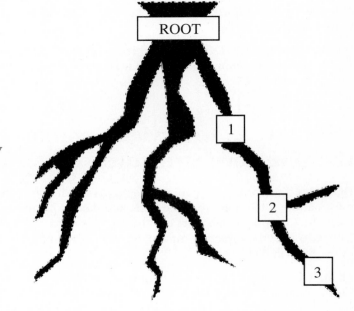

Fig. 4.1
The DOS directory
structure.

In figure 4.2, the numbered boxes represent directories on branches of the
tree-structured hierarchy.

The tree-structure analogy loses some of its neatness when it expands to cover
the capabilities of the hierarchical directory structure. Any directory except the
root can have as many subdirectories as space on the disk permits. Depending
on the disk drive, the root directory can handle a preset number of subdirectories. Hard disks have a typical root directory capacity of 512 entries.
By contrast, 3 1/2-inch 720K and 1.4M floppies can hold, respectively, 112
and 224 entries.

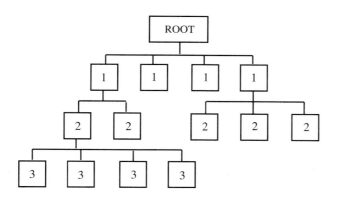

Fig. 4.2
Another depiction
of the DOS direc-
tory structure.

4

When you format a disk, DOS creates a main directory for that disk, called the *root directory*, which is designated by the backslash (\). The root directory is the default directory until you change to another directory. You cannot delete the root directory.

A *subdirectory* is any directory except the root directory. A subdirectory can contain data files as well as other, lower subdirectories. Subdirectory names must conform to the naming rules for DOS files, but subdirectories normally do not have extensions. By naming subdirectories for the type of files they contain, you can remember the type of files each subdirectory contains.

The terms *directory* and *subdirectory* are frequently used interchangeably. A subdirectory of the root can have its own subdirectories. By naming the branches, you can describe where you are working in the tree structure. You simply start at the root and name each branch leading to your current branch.

Disk directories also are frequently called *parent* and *child* directories. You can compare this structure to that of a diversified corporation with numerous subsidiaries (see fig. 4.3). Each child of the parent can have children of its own. In the directory hierarchy, each directory's parent is the directory just above it.

Directories do not share information about their contents with other directories. In a way, each subdirectory acts as a disk within a bigger disk. This idea of privacy extends to the DOS commands that you issue. Unless you specify otherwise, DOS commands act on the contents of the current directory and leave other directories undisturbed.

When you issue a command that specifies a file but not a directory, DOS looks for that file in the default, or current, directory.

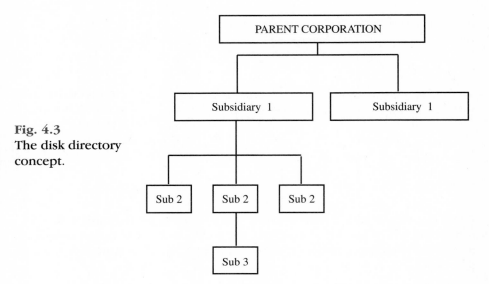

Fig. 4.3
The disk directory
concept.

4

Objective 3: To Understand DOS Paths

You can access any point in the tree structure and remain at your current
directory.

Before DOS can locate a file in the tree structure, DOS must know where to
find the file. The *directory specifier* tells DOS which directory holds a certain
file. In order to find a file, DOS must know the drive you want to use, the
directory name, and the name of the file. In the command line, you type the
disk drive, the directory name, and then the file name. DOS uses this information to find and act on the file. To list the files in the DOS subdirectory of
drive C, for example, you enter the command **DIR C:\DOS**.

Path Names

You can compare DOS to a corporate empire with an extremely strict order of
command. All communications must "go through channels." If, for example, a
subsidiary at level 3 wants to communicate with the parent corporation, the
message must go through subsidiary 2 and subsidiary 1. In DOS, this routing
is called a *path* (refer to fig. 4.3).

A *path name* is a chain of directory names that tells DOS how to find the file that you want. The DOS prompt requires you to build complete path names. When you use the Shell, the paths are supplied visually in the Directory Tree area. If you understand paths now, you can use the command line easily later.

To create a path name, you type the drive name followed by a colon (:), a subdirectory name (or sequence of subdirectory names), and the file name. Make sure that you separate subdirectory names with a backslash (\). In symbolic notation, the path name looks like the following:

d:\directory\directory. . .\ filename.ext

In this notation, *d:* is the drive letter. If you omit the drive specifier, DOS uses the logged drive as the default drive. *directory\directory. . .* names the directories that you want to search. The ellipsis (...) indicates that you can add other directories to the specifier list. If you omit the directory specifier from the path name, DOS assumes that you want to use the current directory.

filename.ext is the name of the file. Notice that you use a backslash (\) to separate directory names from the file name. The path name fully describes to DOS where to direct its search for the file.

Figure 4.4 shows a simple directory setup to illustrate directory paths in DOS. Each subdirectory in this sample is a subdirectory of the root directory. The subdirectory LOTUS contains a data file called MYFILE.123. Another subdirectory, TEMP, contains a file named TAXFORMS.DOC.

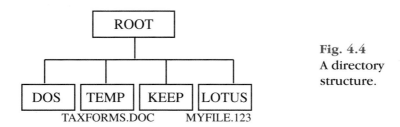

Fig. 4.4
A directory
structure.

MYFILE.123 is a data file in the LOTUS subdirectory. The complete path name for this file is the chain of directories that tells DOS how to find MYFILE.123. In this case, the chain consists of just two directories: the root (\) and LOTUS. Figure 4.5 illustrates the path names for MYFILE.123 and TAXFORMS.DOC.

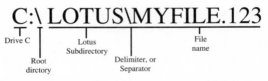

Fig. 4.5
Understanding
path names.

Exercise 3.1: Viewing the Root Directory

In this exercise, you view the root directory and determine how many
subdirectories exist on your computer.

1. Boot your computer if you have not already done so.

2. At the C:\ prompt, type **dir c:\ /p** and press ⏎Enter.

    This command tells the computer to list, one page at a time, all files in
    the root directory of drive C.

3. Look at the listing. How many subdirectories are listed? ***Hint:*** A
    directory name, in this DIR format, is indicated by <DIR> in the space
    where the file name extension normally appears.

4. Look at the listing again (reissue the command in step 2, if neces-
    sary), and determine how many files (not directories) are in the root
    directory. Advanced hard disk users usually attempt to keep only
    AUTOEXEC.BAT, CONFIG.SYS, and small batch files in the root
    directory.

Starting the Search

When you type a path name, DOS searches in the first specified directory and
then passes through the other specified directory branches to the file. The
root directory, from which all directories grow, has no name and is represented by the backslash (\). If you want the search path to start at the root
directory, begin the directory specification with \. DOS begins its search for
the file in the root and follows the subdirectory chain that you include in the
command.

Suppose that you want to see the directory listing for a budget file that you created using Lotus 1-2-3. You may type a DIR command similar to the following to give DOS a path:

dir c:\lotus\data\budget.wk1

DOS searches drive C, beginning with the root directory, proceeds to the LOTUS subdirectory, and then arrives at the DATA subdirectory.

If you omit the root name designator, DOS searches for files in the current directory, which DOS uses as the default starting point for its search. If the current directory doesn't lead to the subdirectory that contains the file, you see a File not found error message.

When the current directory contains the subdirectory, you do not have to type all the directory names in the path. For the preceding command, for example, if the current directory is C:\LOTUS, you can view the listing for the budget file by typing the following command at the DOS prompt:

dir data\budget.wk1

Determining the Path

The AUTOEXEC.BAT file, which helps set your working environment when you boot the computer, contains a path line that tells the computer the order in which directories should be searched when you request a file.

The symbolic syntax for the PATH command is as follows:

PATH *d1:\path1;d2:\path2;d3:\path3...*

Exercise 3.2: Viewing the Path

To see your path on your computer, follow these steps:

1. Boot the computer, if necessary.
2. At the C:\ prompt, type **path**, and press ⏎Enter.
3. A path statement appears that looks similar to the following:

   ```
   path=c:\;c:\wp\;c:\dos\;c:\123\;c:\dbase
   ```

   This path tells the computer to look first in the root directory, then in the WP directory, then in the DOS directory, then in the 123 directory, and finally the DBASE directory. If DOS cannot locate the file you have requested in any of these directories, it displays a File not found error message.

4. Examine your path, and determine which directories your computer is set up to search.

Objective 4: To Understand and Explore Subdirectories

In this section, you explore further the organization of your hard drive by using the TREE command, which provides a structured view of the hard drive.

The symbolic syntax for the TREE command is as follows:

TREE *d:* */F/A*

The /F switch lists all the files in each directory. The /A switch uses standard characters rather than graphic lines to display the directory structure.

Exercise 4.1: Using TREE To Display All Directories

To use the TREE command to display your directories, follow these steps:

1. Boot your computer, if necessary.
2. At the C:\> prompt, type tree, and press ⏎Enter.

   Items in the root directory are displayed to the far left of the directory listing. Items in subdirectories are connected with horizontal lines off the root directory items.
3. Determine how many subdirectory levels you have under your root directory.

The remainder of this section explains the purpose of some of the directories on your hard disk. Because the configuration of each hard disk is unique, your hard disk probably does not contain all the subdirectories described. Nevertheless, you can use the explanations given to determine whether you should have a subdirectory for a particular purpose.

The Root Directory

DOS creates the root directory, the top directory in the inverted tree, but you control which files to include in it. As a general rule, avoid cluttering the root directory of a hard disk with files.

Because the root is the default directory for DOS when you boot your system, you must include COMMAND.COM in the root directory of your first hard disk and any floppy disk that you use to boot your computer. DOS expects to find COMMAND.COM in the current directory when you boot. If DOS cannot load COMMAND.COM, it cannot communicate with you; it simply warns you that it cannot find the command interpreter.

In addition to COMMAND.COM, the root directory probably contains the AUTOEXEC.BAT and CONFIG.SYS files. DOS uses these files when you boot the computer.

The root directory should contain few, if any, additional files. Almost all other files should be placed in an appropriate subdirectory.

The \DOS Directory

When you install DOS on your hard disk, the installation procedure creates the \DOS directory and copies the DOS files into this directory. Never place files other than DOS files in the \DOS directory. Doing so increases the risk of accidentally destroying important files when you upgrade to a new version of DOS.

The \BATCH Directory

Batch files are text files that contain DOS commands and execute programs. You can place many commands in one batch file. Even if you use the Shell to execute commands and programs, you still may use batch files. Most users keep their batch files in a separate \BATCH directory.

The \UTIL or \UTILITY Directory

Just as you keep all your DOS files in the \DOS directory, you may want to keep utility programs in their own directory. Most people accumulate a variety of small utility programs such as print spoolers, mouse drivers, file compression utilities, and disk utilities and keep them in the \UTIL or \UTILITY directory. Placing these utility programs in the \DOS directory is not advisable. If you upgrade your version of DOS later, these programs may no longer be available to you. Never place any files in the \DOS directory except DOS files.

The \DATA Directory

You probably use your personal computer for many different purposes and have many different data files. Some users create a data subdirectory under each program directory. If they have a \LOTUS directory for the 1-2-3 program files, for example, they create a \LOTUS\DATA or \LOTUS\FILES directory for worksheet files.

This structure works, but a better system is to create a \DATA directory with subdirectories for each project or application. Store all files related to your taxes, for example, in the \DATA\TAXES directory. If you take a finance class, all your homework assignments may go in the \DATA\FINCLASS directory. The \DATA\FINCLASS directory may contain worksheet and word processing files, but they are all related to one activity.

Using a \DATA directory enables you to back up your data easily without making a backup copy of your programs.

The \TEMP Directory

Many users find that they need a directory to store temporary files, making a directory named \TEMP useful. You can copy files to \TEMP as a temporary storage place until you copy the files to a more appropriate directory.

If you have a single-floppy, low-memory system, a \TEMP directory also enables you to make copies of floppy disks by copying files from the source disk to the \TEMP directory and then back to the destination disk. With this method, you don't have to keep swapping disks in and out of the single floppy drive.

Do not use the \TEMP directory as a permanent home for files, however. You should be able to erase all the files in this directory periodically, keeping it empty for later use.

If you use Microsoft Windows, be careful when using a TEMP directory. When you install Windows, a C:\WINDOWS\TEMP directory is created to store temporary files. Because Windows uses these files, you should never delete any files from the Windows temporary directory while Windows is running. You can, however, create the directory C:\TEMP. Although the directories are both named TEMP, the complete paths are different, and DOS knows that the two directories are unrelated.

The \MISC or \KEEP Directory

You may have files in different directories that are no longer active but that you believe you may still need. Inactive files in a directory tend to increase clutter and make sorting through the directory confusing. With a \MISC or a \KEEP directory, you have an easily remembered home for inactive files. Of course, delete only those files that are clearly of no more use to you.

Applications Software Directories

Many applications packages create directories when you install them on your hard disk. If a program does not create a directory, you should create one with a name that suggests the software name. You may, for example, name your spreadsheet directory LOTUS. You can then copy the 1-2-3 package files to the directory.

If you work with multiple versions of the same program, you can name a subdirectory 123R22 for the 1-2-3 Release 2.2 program files and 123R3 for the 1-2-3 Release 3 program files. If you have enough disk space, keep the old and the new versions of a program for a while in case you have problems with the new version.

Objective 5: To Manage Directories on Your Hard Drive

The examples presented so far in this chapter provide information on the structure of hierarchical directories. The commands listed in this section relate to the maintenance and use of your directory system. With directory commands, you can customize your file system and navigate through it.

Warning: Never change the directory structure (add or delete directories and/ or files) of a computer that you do not own or directly manage. The directory structure on most machines is designed to meet specific software and user specifications. If a directory is deleted or renamed, certain software products may no longer work. For this reason, the exercises in this section direct you to create, delete, and copy files to and from a floppy disk. Please check to make sure that your working directory is A at all times.

Creating a New Directory

The symbolic syntax for the DOS command to make a new directory is as follows:

MKDIR *d:directory name*

or

MD *d:directory name*

This book uses the abbreviated MD command for all make directory exercises.

Exercise 5.1: Making a WORK Directory on Your Floppy Disk

To create a directory on your floppy disk, follow these steps:

1. Boot your computer, if necessary.
2. Insert a previously formatted floppy disk into drive A, or format a floppy disk if necessary.
3. Change the logged disk drive to drive A by typing **a:** and pressing ⏎Enter.

   The DOS prompt should now be A:\>.
4. Make a directory named WORK on drive A by typing **md a:\work** and pressing ⏎Enter.
5. DOS doesn't confirm that it has executed your command, so you need to check. Type **dir a:** and press ⏎Enter to make sure that the entry WORK <DIR> appears in the directory listing.

Changing the Current Directory

In the preceding exercise, you created a new directory. You probably noticed, however, that DOS did not automatically move you into that new directory. To change directories, you must issue a separate command.

The symbolic syntax for the command to change directories is as follows:

CHDIR *d:path*

or

CD *d:path*

Again, the abbreviated version of the command is used here.

Exercise 5.2: Changing into the WORK Directory

To change to another directory, follow these steps:

1. Boot the computer, if necessary.
2. Insert the floppy disk you used in Exercise 5.1 into drive A.
3. Type **a:** and press ⏎Enter to change to drive A.
   The DOS prompt should now be A:\>.
4. To change into the WORK directory, type **cd a:\work**, and press ⏎Enter.
   The DOS prompt should now be A:\WORK>.
5. Type **dir**, and press ⏎Enter to view the WORK subdirectory.
6. Continue with Exercise 5.3.

Exercise 5.3: Changing to the Root Directory

After you have changed to a named directory, the concept of changing directories seems relatively straightforward. Nevertheless, users often get stuck when they want to change back to the one allowable unnamed directory—the root directory.

1. Make sure that the DOS prompt is A:\WORK>.
2. To change back to the root directory, type **cd a:** and press ⏎Enter. Remember that the root directory doesn't have a given name.

   Because you have not specified a name after the backslash, DOS knows to move to the root directory.

Make sure that you understand the concept of changing directories before you continue; go back and practice these two exercises until you are comfortable with the commands for changing directories.

Removing a Directory

You may find that rearranging your hard drive occasionally can make the disk structure more useful to you. During this rearrangement, you may find that you don't need a directory. Before you remove a directory, however, you must first delete all the subdirectories and files from the unwanted directory.

The symbolic syntax for removing a directory is as follows:

> **RMDIR** *d:path*

or

> **RD** *d:path*

Exercise 5.4: Removing the WORK Directory

To remove the directory you created in a preceding exercise, follow these steps:

1. Boot your computer, if necessary.
2. Insert the floppy disk containing the WORK directory into drive A.
3. Type **dir a:\work** and press ⏎Enter to make sure that the WORK directory is empty.

    The WORK directory should have two directories: the parent and the child directory (listed as . and ..). Don't worry about these.
4. To remove the directory, type **rd a:\work** and press ⏎Enter.
5. Type **dir a:** and press ⏎Enter to make sure that the directory has been deleted.

Chapter Summary

This chapter emphasizes the importance of file organization and hard disk management. A little forethought and planning go a long way in making the large storage capacities of modern computers work for you rather than against you. Understanding how DOS uses the set path to search for files will help you know what to do when you get a File not found error message. Knowing how to navigate between directories, and to make, change, and remove directories gives you the power to organize your computer data in a way that is comfortable and efficient for you.

Testing Your Knowledge

Next Thursday
Thursday
02/02/95

True/False Questions

T 1. DOS directories are structured in a hierarchical format.

T 2. Files may have the same name if they are stored in separate directories.

F 3. DOS searches for a requested file according to the alphabetical order of the directories on the disk.

F 4. A good method for managing files on the hard drive is to place all work files in the root directory, where they are easily accessible.

F 5. The CD command enables you to change drives.

Multiple Choice Questions

1. Hard disk file management in DOS is analogous to
   - A. a wastebasket.
   - B. a desktop.
   - ✓ C. a filing cabinet.
   - D. a bookshelf.
   - E. a closet.

2. The _____TREE_____ command shows the entire disk structure in a graphical format.
   - A. DIR
   - ✓ B. TREE
   - C. SHOW
   - D. DIR/GRAPHICAL
   - E. STRUCT

3. Which of the following files contains the path statement?
   - A. COMMAND.COM
   - B. CONFIG.SYS
   - C. PATH.SYS
   - ✓ D. AUTOEXEC.BAT
   - E. FORMAT.COM

4. To change directories from the FILES subdirectory to the root direc-
   tory on the hard drive, you must use the command
   - A. /FILES /
   - B. CD C:\
   - C. CD
   - D. MD ROOT
   - E. CD C:\ROOT

5. To change drives from drive C to drive A, use the command
   - A. CD C: A:
   - B. MD C: A:
   - C. CD A: C:
   - D. A:
   - E. C: to A:

Fill-in-the-Blank Questions

1. The ___*path* bAck slash___ lists the search structure on a given drive.
2. The ___RD___ deletes a directory from the drive.
3. The ___mD___ creates a directory on the drive.
4. In order to remove a directory, the directory's contents must be
   ___RD A:\Name: — empty___
5. The root directory must contain ___Command.com___ files.

Review: Short Projects

1. Creating Directory Structures

   Write the commands necessary to create the following directory
   structure:

C:\> A:\> USER
C:\ UTIL\TXT

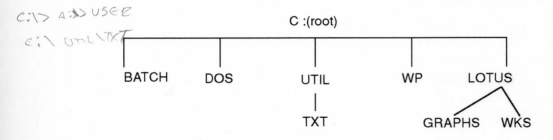

110 DOS

2. Drawing a Hard Disk Organizational Structure

   Draw a hard disk organizational structure to organize your files in an efficient manner.

3. Examining the Path of a Hard Drive

   Visit a public computing site on your campus, and examine the path on one of the hard drives. Diagram the route DOS takes when searching for files.

Review: Long Projects

4

1. Creating Floppy Disk Directory Structures

   Create the following directory structure on a floppy disk:

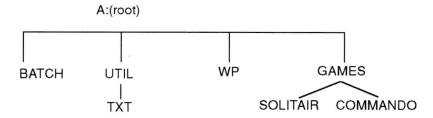

2. Deleting a Subdirectory

   Using the directory structure you created in Project 1, delete the TXT subdirectory in the UTIL directory.

3. Writing DOS Search Statements

   Write path statements that cause DOS to search in the following directories:

   A. A:\

   B. C:\DOS

   C. C:\UTIL

   D. C:\

Maintaining Files

5

In this chapter, you reexamine the concept of paths by working with files in various directories. You also consider some important issues regarding safe storage of files and measures that can be taken both to prevent and to recover from file loss. This chapter tells you how to copy disks, copy and move files, erase unneeded files, and rename existing files. The chapter also deals with the important file-maintenance procedures for archiving your files for safekeeping. This process, known as *backup*, is one that you should get into the habit of doing on a regular basis.

Objectives

1. To Label Disks and Files
2. To Use Absolute and Relative Paths
3. To View Files in Other Directories
4. To Copy Files
5. To Delete Files
6. To Use DISKCOPY To Copy Disks
7. To Avoid Data Loss
8. To Recover Deleted Files

Key Terms in This Chapter	
Source	The disk or file from which you are copying or moving.
Destination	The disk or file to which you are copying or moving.
Target	Same as *destination*.
Current directory	The directory that DOS uses as the default directory. The current directory is highlighted in the Directory Tree area in the DOS Shell.
Overwrite	Writing new information over old information in a disk file.
Surge suppressor	A protective device inserted between a power outlet and a computer's power plug. Surge suppressors help block power surges that often damage computer circuits.
Static electricity	An electrical charge that builds on an object and can discharge when another object is touched. Electronic circuits are easily damaged by static electricity discharges.
Self-parking heads	Heads on most newer disk drives that move to a position on the disk in which there is no data. These heads protect the disk from data loss when it is moved.
Head-parking program	A program that parks heads on a disk that does not have self-parking heads.
Delete-tracking file	A special file, created by the MIRROR command, used by DOS to make it easier and more reliable to undelete files.
Backup	To copy files from the hard disk to floppy disks.
Restore	To copy backup copies of files on floppy disk to a hard disk.

Objective 1: To Label Disks and Files

Disk files are the primary storage place for data and programs. A knowledge of how to manage these files is essential. If you want to be in control of your work, you must be in control of your files.

When you work with floppy disks, always keep the labels on your disks current. Use a felt-tip pen, and indicate the contents on the disk label as you work. Disks not labeled or labeled incorrectly are an invitation to lost data. If you do not label disks, you may mistake them for blank, unformatted disks.

Never use a ball-point pen to write on a label that has been placed on a floppy disk. The jacket may not keep the pen point from damaging the magnetic media and harming the disk.

5

Objective 2: To Use Absolute and Relative Paths

You had some practice copying files in Chapter 3. Now that you know about directories and paths, however, you may want to look at the copying process again.

In Chapter 4, each time you were instructed to type a command, the command included the full path for the location of the file. When drive A was the default drive, for example, and you were asked to run a directory of the files on the disk in drive A, you typed *dir a:*. This method is called *giving the full path*. Providing the full path in a command is the safest method of issuing DOS commands, because it leaves little room for error.

In computing, you deal with absolute and relative locations all the time. This concept comes up in spreadsheets and databases, as well as general file management, with which you are dealing here. Giving the full path in a DOS command is more generally called giving an *absolute reference*. The absolute reference tells the computer exactly where to locate something, regardless of which drive or directory is the current default. The alternative, giving a *relative address*, relies on your current drive and directory location. If the default drive is A, for example, and you want to get a directory of the files on the disk in drive A, the relative command you type is DIR. In this case, the computer assumes that you want a directory of the disk in drive A, because you haven't specified a different directory. Although relative addresses are shorter, they can get you into trouble.

As an analogy, if a stranger comes to the door of your house and asks, "Do you know where Bob Smith and his family live?" you can reply in a couple of ways. If you know the Smith's address, you say, "The Smiths live at 5229 Greenbriar Trail." This answer tells the inquirer precisely where the Smiths live. Providing the street address means that your instructions are not dependent on your current location. Providing the exact street address is analogous to giving the full path in a DOS command.

An alternative reply to the inquiry about the Smiths may be "Oh, the Smiths live two doors down, on the left side of the street." This reply also tells the inquirer where the Smiths live, but the directions depend on the current location. In other words, if the stranger went back to a gas station before going to the Smith's house, he would first have to locate your house and then count two doors down. This response is analogous to giving a relative address, because the directions are relative to the current location of your house.

Objective 3: To View Files in Other Directories

Your instructor has stored work files for you in a directory of the hard drive called DOS_WORK. You will spend some time looking at these files, using variations of the DIR command.

Exercise 3.1: Using the DIR Command

In this exercise, you practice using the full path and relative path with the DIR commands. Although this book gives you each command to enter, try to challenge yourself to figure out the command on your own. Make sure that you're not just typing the printed commands. You will learn only if you constantly ask yourself, "Do I really understand why this works this way?"

1. Boot your computer, if necessary, and insert a formatted floppy disk into drive A.

2. Check to see whether the DOS_WORK directory is on your hard drive by typing **dir c:\dos_work** and pressing ⏎Enter. This command is the full path command for this task.

3. Use the relative path command to look for the DOS_WORK directory by typing **dir dos_work** and pressing ⏎Enter.

    Notice that the results of the commands in steps 2 and 3 are the same.

4. To change the default drive to drive A, type **a:** and press ⏎Enter.

Which command will you have to use to see the DOS_WORK directory? Will you use the one given in step 2 or the one given in step 3? Why?

Objective 4: To Copy Files

You learned in previous chapters that the process of copying a file involves issuing the COPY command with a source file location and name and a destination file location and optional name. You also learned that the result of a copy procedure is that one or more files are *duplicated*—the same file exists in two different places.

5

Copying Files across Directories

Copying a file or a group of files from one directory to another directory is just like copying a file or a group of files from one disk to another disk. The only difference is that you must specify the full path for the source and destination files.

Exercise 4.1: Copying Files to and from Different Directories

In this exercise, you use work files distributed by your instructor. You will need a formatted floppy disk for this exercise.

1. Boot your computer, if necessary.
2. On your floppy disk, create a temporary directory. You will copy files into this directory and use them for the remainder of this chapter. Type **md a:\tmp** and press ⏎Enter.
3. Copy all files that begin with the letter *C* from the DOS_WORK directory on the hard drive to the \TMP directory on drive A. You can do this in one command. Type

    copy c:\dos_work\c*.* a:\tmp

   Press ⏎Enter.

   Notice that with this command, you must give the full path for both the source and destination files. How could you have shortened this command?

4. Change to the C:\DOS_WORK directory by typing

    cd c:\dos_work

    Press ⏎Enter.

5. Now copy all files that end with TXT from the DOS_WORK directory to the \TMP directory on drive A. How will the command for this task differ from the command in step 3? Type

    copy *.txt a:\tmp

    Press ⏎Enter.

Because the default drive is drive C and the default directory is DOS_WORK, you don't need to state that location in the COPY command. If the default drive were drive A and the default directory were \TMP, what would the command for this task look like?

By now you should start to see the utility of both full-path DOS commands and relative-path DOS commands. Relative-path DOS commands enable you to type fewer keystrokes—but assume that you know where you are. Full-path commands, on the other hand, require more keystrokes but work correctly regardless of your current location (disk drive or directory). Most computer users employ a combination of mixed-path commands—they use both full and relative paths.

Copying Files in the Same Disk Directory

One rule of DOS is that no two files in a single directory can have the same name. Why doesn't DOS allow this? Think of a directory on a disk as a house on a block. Suppose that two people with the name Robert Smith live in the same house. When you want one of them, you call, "Hey, Robert Smith!" Both men come running to you. To avoid confusion, you might call one Bob Smith and the other Rob Smith. The same dilemma faces DOS. You cannot have two files named MEMO1229.DOC, for example, in the same directory. You can have, however, one file named MEMO1229.DOC and another file named MEMO1229.TXT. Although the root name is the same, MEMO1229, the file extensions are different—DOC and TXT.

Perhaps you have just created a file called EXPJAN93.WK1 containing your expenses for January. Now you are ready to begin creating a file with your expenses for February. Suppose that your expenses for each month are quite similar. Rather than create a new file, you copy the file EXPJAN93.WK1, calling the new file EXPFEB93.WK1. You then can edit EXPFEB93.WK1 for the correct February expenses.

Exercise 4.2: Copying a Memo

In this exercise, you practice copying files. Follow these steps:

1. Boot the computer, if necessary, and insert the floppy disk containing the \TMP directory into drive A.

2. Copy the file MEMO.TXT from the DOS_WORK directory on the hard disk to the \TMP directory on your floppy disk in drive A. Refer to Exercise 4.1 if you need help doing this.

3. Make the default drive A and the default directory \TMP. Again, refer to Exercise 4.1 for help.

4. Use the TYPE command to view the contents of MEMO.TXT. Type

   type memo.txt

   Press ⏎Enter .

   If you get the error message File not found, you need to make sure that your default drive is drive A and your default directory is \TMP.

   The file MEMO.TXT is a *boilerplate* memo; that is, the format exists for a memo, but none of the real information is there for an actual memo. Boilerplate documents are convenient because you don't have to waste time figuring out the placement of things like *TO:* and *FROM:* each time you create a memo. However, you have to remember to keep the original file, MEMO.TXT, unchanged. For this reason, you must make a copy of MEMO.TXT but rename the copy so that you don't alter the boilerplate file.

5. Copy the file MEMO.TXT, naming the destination copy QUE.MEM. Type

   copy memo.txt que.mem

   Press ⏎Enter .

   Notice that you are using the relative path in this command. What would the full path for this command look like?

6. Run a directory of the A:\TMP directory to make sure that the two memo files are there.

Objective 5: To Delete Files

When you no longer need a file, you can remove the file from the disk. Erasing old files that you no longer use is good computer housekeeping. Free space on disks—especially hard disks—becomes scarce if you do not erase un-needed files.

Exercise 5.1: Deleting Unnecessary Files

In this exercise, you delete all files beginning with the letter *C* from the \TMP directory on the floppy disk.

1. Boot the computer, if necessary, and insert the floppy disk with the \TMP directory on it.

2. Delete the file CAT.TXT from the \TMP directory of the floppy disk. Type

    del a:\tmp\cat.txt

    Press (⏎Enter).

3. Change directories so that the \TMP directory is the default directory. Delete the file CAT.BAK by using the relative-path command.

4. Using the full-path command, delete all files beginning with the letter *C* from the \TMP directory of drive A. Remember that you can do this in one command by using wild cards.

Objective 6: To Use DISKCOPY To Copy Disks

You have learned how to copy one or more files on a disk. You can also duplicate all the files and directories from one floppy disk to another with the DISKCOPY command.

The syntax for DISKCOPY is

   DISKCOPY *sd: dd: /1 /V*

where *sd* stands for *source disk* and *dd* stands for *destination disk*. The */1* switch copies only one side of a disk and is rarely used with modern double-sided disks. The */V* is the verify switch and is used to do a final comparison of the two disks after all information is copied. Using the /V switch is always a good idea.

When you use DISKCOPY, you make an exact copy of another disk. The input (or source) disk is read and then the data is written to another disk, the destination disk. *Any files on the destination disk are lost.* When you choose DISKCOPY, an actual DOS utility is executed. This utility, called DISKCOPY, is an external DOS program.

DISKCOPY is good to use when you want to make backup copies of program disks. You then can store the original disks in a safe place.

The source and destination disks must be the same size and capacity. You cannot use DISKCOPY to duplicate a 5 1/4-inch disk on to a 3 1/2-inch disk or vice versa. DISKCOPY creates an exact duplicate of another disk.

If you have two disk drives that are the same size, DISKCOPY is a breeze. Simply insert your source disk into drive A, the source drive. Insert the destination disk into drive B, the destination drive.

If you have only one disk drive to use for disk duplication (which is what most computers have), you must use the same disk drive for both the source and the destination. DOS tells you when to insert the source disk and when to insert the destination disk as DOS makes the duplicate. The following exercise is based on the assumption that you have only one disk drive for this operation.

Exercise 6.1: Making a Backup of Your Floppy Disk

You need the formatted floppy disk with which you have been working for this chapter, as well as a new, unformatted floppy disk.

1. Boot the computer, if necessary.
2. On the work disk that you have been using throughout this chapter, write the word *SOURCE* on the outside disk label with a felt-tip pen.
3. Place the sticky label on the blank, unformatted disk, and write the word *DESTINATION* on the label. This labeling will keep you from getting the source and destination disks mixed up during the diskcopy process.
4. If you have only one floppy drive, skip to step 5 of this exercise. If you have two floppy drives of the same size, insert the disk labeled "SOURCE" into drive A and the disk labeled "DESTINATION" into drive B. Type the command

   diskcopy a: b: /v

   Press ⏎Enter.
5. If you have only one floppy drive, insert the disk labeled "SOURCE" into drive A. Then type the command

   diskcopy a: a: /v

   Press ⏎Enter.

Watch the computer screen carefully. When prompted, remove the source disk and replace it with the destination disk. You will need to swap disks approximately 20 times for a high-density disk. Make sure that you do not get the disks confused—always check the disk label to ensure that you are using the disk for which the computer has prompted you.

Objective 7: To Avoid Data Loss

As you use your computer, you create files, many of which contain valuable information. Because you are very careful and today's computers are very reliable, you may be tempted to trust that these files will be there when you need them. However, as an old computer saying goes, "There are two kinds of computer users: those who have lost files, and those who are going to lose files." You need to know how to avoid losing your valuable data.

Taking Preventive Measures To Avoid Data Loss

The first way to avoid data loss is to prevent it before it happens. Unfortunately, no matter what you do, it is impossible to guarantee that you will never lose data. You can minimize the risk, however, with preventive measures.

Preventing Mistakes

Each time you use your computer, you gain more and more experience. You create, copy, and erase more files. You also become a little less careful when you copy and erase files. One difference between a novice and an expert is that the expert makes many more mistakes.

Commands such as COPY, ERASE, and FORMAT perform their jobs without regard to your intentions. DOS does not know when a technically correct command will produce an unwanted effect. You therefore should always study the commands that you enter before you execute them.

The most common errors that cause data loss do not occur in DOS, however. Most people spend much more time using applications programs than they spend using DOS. You can lose data in your spreadsheet, word processor, or database management system in many ways. You can issue a command that erases a large part of your forecasting spreadsheet, for example. Then, the next time you retrieve your spreadsheet, you may find that some of your data is gone.

Just as you must exercise care when you use DOS commands, you must be equally careful when you use commands in your applications. Each application has different commands; therefore, make sure that you have copies of all your important files before you change them.

Preventing Software Failures

Each software program you buy is a set of instructions for the microprocessor. Some software packages have mistakes called *bugs*. Software bugs are usually minor and rarely cause more than keyboard lockups or jumbled displays.

Utility programs, such as disk caches and partition utilities, however, can interfere with complex programs, such as Windows. A faulty utility program can wipe out an entire hard disk. This occurrence is rare, but that fact is little consolation if you lose all your data.

Bugs usually occur in the first releases of programs. Most software contains version numbers. Version 1 or 1.0 is the first version of a program. Version 1.1 is a minor upgrade, and Version 2.0 is a major upgrade. Many people try to avoid Version 1 of any program. Some people avoid Version x.0 of any program (such as 2.0 or 3.0), because a major upgrade is also likely to have bugs.

Perhaps the best way to avoid software errors is to talk to other people about a program before you buy or use it. Talk to friends and coworkers. You can meet people at computer user group meetings and find out about their experiences with the program. If the program gets good reviews, the chances of serious software errors are minimized.

Another type of software problem that can cause serious data loss is a *virus*. A computer virus is a set of computer instructions, hidden inside a program, that can take over your computer and destroy all your programs and data files. Viruses are the work of computer vandals who destroy the property of others for "fun."

A virus rarely infects commercial software. Viruses usually are found in free software distributed through electronic bulletin board systems (BBSs) and passed around on floppy disks. Operators of bulletin board systems work very hard to avoid viruses, but the risk is not completely eliminated.

To practice "safe computing," never use a program that you get from someone you do not know. Before you use any program, talk with others who have used the program and make sure that they have had no problems. Also make sure that the date and file size of both versions of the program are identical. Two "copies" of what should be the same program with different file sizes is a clue that the larger one is infected with a virus.

You can obtain special programs that test your system for viruses and remove the viruses from your software if any are found. These programs are available both commercially and from PC user groups. Even an experienced computer user calls for expert help if a computer virus is suspected. Membership in a local PC user group is a good way to have access to an expert if you need one.

Preventing Hardware Failures

Today's personal computers are reliable and economical data-processing machines. The latest generation of PCs does the work of mainframe computers, which, a decade ago, only a fortunate few could access. As is true of any machine, however, computer components can break down.

Computers contain thousands of integrated circuits. Under ideal conditions, most of these circuits can last a century or more. Disk drives incorporate precise moving parts with critical alignments. Although disk drives are very reliable, most common hardware failures occur on disk drives.

Computer hardware is vulnerable to a number of physical threats. These threats include humidity, static electricity, excessive heat, and erratic electrical power. By following the precautions presented in this section, you reduce the odds of losing time and information because of hardware failure.

You will experience hardware failure sometime. The failure may be due to the environment or to simple bad luck. A memory chip may fail for no apparent reason. A disk drive may fail in a number of different ways. The result is that you lose access to your data.

You cannot overcome bad luck, but you can be vigilant about your computer's environment. A power strip with a built-in surge protector is a good start. If your power flutters and lights flicker, you probably need a line-voltage regulator. Make sure that no electrical appliances near your computer pollute your power source. Connect your computer equipment to power sources not shared by copiers, TVs, fans, or any other electrical equipment that contains a motor or uses a surge of power when it is turned on.

Is the fan on the back of your computer choked with dust? Clean the air vents and check that your computer has room to breathe. Your computer can become erratic when the temperature climbs. Circuits are not reliable when they overheat and can cause jumbled data. To make sure that your computer can breathe, keep it cool.

Use a soft blind-cleaner attachment on your vacuum cleaner to remove the dust build-up from your computer's breathing system. Always be sure to turn off *and* unplug your computer before vacuuming around it. Frequencies from the vacuum and a powered-on computer may match, and each appliance may short out the other.

If the outside of your computer is dusty, the inside may be full of dust, too. Many computer stores offer "tune-up" specials to clean the inside of your system unit and disk drives. You should have this done about once a year, or more often if you work in a smoky or dusty environment.

Your body generates static electricity when humidity is low, when you wear synthetic fabrics, or when you walk across carpet. Static electricity appears harmless, but electronics are very sensitive to it.

Just touching the keyboard while carrying a static charge can send an electrical shudder through your computer, causing data loss or circuit failure. Fortunately, you can avoid static problems by touching your grounded system unit's cabinet before touching the keyboard. If static electricity is a serious problem for you, ask your dealer about antistatic products.

Moving or shaking your computer can damage the disk drives. Never move your computer while the power is on and the hard disk is spinning. While a hard disk is on, the heads float a fraction of an inch above the disks.

When you turn off your computer, the hard disk stops spinning and the heads settle onto the disks. If you move your computer with the power off, the heads can move or bounce on the disks and damage the surface of the disk. Most hard disks have self-parking heads to guard against disk damage. When you turn off your computer, the heads move to a part of the disk that is not used for data. Even if the heads damage the disk surface, no data is lost. If your hard disk does not have self-parking heads, you can use the head-parking program that came with your hard disk to park the heads manually. Always run the park program before you move your computer.

The heads on floppy disks can be damaged when you move your computer as well. To protect the heads, insert a floppy disk that does not contain any data and close the drive door before you move your computer.

Stopping small hardware problems before they become big problems takes a little planning and forethought. Table 5.1 lists a few simple, yet successful, preventive solutions.

Table 5.1 Hardware Problems and Remedies	
Problem	*Remedy*
Damaged disks	Don't move the computer while the disk is running.
Damaged floppy disks	Don't leave disks where they may be warped by the sun. Use protective covers. Avoid spilling liquids on disks. Store disks in a safe place. Avoid magnetic fields from appliances (TVs, microwave ovens, and so on).
Damaged hard disks	Park the heads before you move the computer.
Overheating	Clean clogged air vents and make sure that objects are not blocking vents. Work in an air-conditioned room in the summer, if possible.
Static electricity	Use an antistatic spray on the carpet around your computer desk. Commercial products are available, or you can make your own spray by mixing one part liquid laundry softener with three parts water in a spray bottle. You can also purchase antistatic floor mats and touch pads if your static electricity problems are severe.

Objective 8: To Recover Deleted Files

With earlier versions of DOS, data recovery was not possible. After you erased a file or formatted a disk, all the data was gone forever. In many cases, however, the data still existed on the disk, but you could not access it.

Many utility programs were written to recover this lost data. Some of the most well-known programs were part of the Norton Utilities, Mace Utilities, and PC Tools Deluxe. Microsoft incorporated some of the utilities from PC Tools Deluxe, Version 6, into DOS 5 to provide recovery for lost data. With these utilities, you can undelete a file that you erased and unformat a disk that you accidentally formatted.

These programs are possible because the ERASE and FORMAT commands do not really remove files from the disk. Instead, these commands clear the information from the file allocation table (FAT) and the directory. The file allocation table tells DOS which areas of the disk contain data and which areas are available. The directory tells DOS information about the file, such as the file name and size.

After you erase a file or format a disk, all the files are still on the disk, but the space is freed by DOS so that the next time you create a file, DOS can write over these files. Therefore, if you want to recover a file, you must do it immediately, before the disk space is used again to write another file.

Some lost data cannot be recovered. If you copy a file to a disk or subdirectory that already contains a file with that name, the old file is overwritten and is lost. If you save a file in an application program with the same name as an existing file, the program may give you a warning message. If you override the warning and save the file, the old file is overwritten and is lost. Your only recourse is to restore a backup copy.

5

Using MIRROR To Prevent Data Loss

The MIRROR command does not recover lost data, but it saves information about your disks to make it easier for DOS to recover lost data. The MIRROR command provides two separate facilities to prevent data loss. You can use MIRROR as follows:

- Make a copy of the file allocation table and directory. If this area of the disk becomes damaged, you can use the copies of these two system areas to recover the information on the disk with the UNFORMAT command.

- Create a delete-tracking file that keeps track of every file you erase. With this file, you can undelete files easily and reliably.

The syntax for the MIRROR command follows:

MIRROR [*drive1:*] [*drive2:...*] [*/tdrive1*] [*/tdrive2...*]

If you specify a *drive1*, such as

MIRROR C:

DOS saves the file allocation table and directory for the specified drive. If you do not specify a drive, DOS saves the file allocation table and directory for the current drive. You can also specify multiple drives.

You should install delete tracking every time you boot your computer. The best way to ensure that you have delete tracking enabled is to put the MIRROR command in your AUTOEXEC.BAT file. The AUTOEXEC.BAT file is explained in Chapter 7.

Exercise 8.1: Running MIRROR

In this exercise, you enable the MIRROR utility, as preparation for future exercises. You need a floppy disk for this exercise.

1. Boot your computer, if necessary, and insert your work disk into drive A.
2. Turn on the MIRROR utility to track the FAT table and record up to 50 file deletions on drive C. Type

    mirror c: /tc-50

    Press ⏎Enter.

Using Undelete To Recover a Deleted File

If you accidentally delete a file or group of files, you may be able to recover the file if you use the UNDELETE command immediately. Remember, when you tell DOS to delete a file, the area on the disk that contains that file is marked as available. The next time you write a file to the disk, DOS may use this part of the disk for the new file, and the deleted file can no longer be recovered.

If you delete a file after you have run MIRROR to install delete tracking, you can easily undelete the file. The syntax for this UNDELETE operation follows:

    UNDELETE *d:path\filename.ext* /LIST

The /LIST switch lists the files in the delete-tracking file but does not undelete any files.

To undelete one or more files, use the UNDELETE command without the /LIST switch. You can also use wild cards (* or ?) to undelete multiple files. The syntax for this UNDELETE operation follows:

    UNDELETE *d:path\filename.ext* /DT /DOS/ALL

The /DT switch tells UNDELETE to use a delete-tracking file created by MIRROR. The switch /DOS tells UNDELETE to ignore the delete-tracking file and use the DOS directory. UNDELETE normally prompts with (Y/N) when undeleting files (see fig. 5.1). The /ALL switch, however, tells UNDELETE to undelete all files without asking.

```
Directory: C:\DATA\DOCS
File Specifications: *.DOC

   Deletion-tracking file contains     1 deleted files.
   Of those,     1 files have all clusters available,
                 0 files have some clusters available,
                 0 files have no clusters available.

   MS-DOS directory contains     1 deleted files.
   Of those,     1 files may be recovered.
Using the deletion-tracking file.

      SCHEDULE DOC     5379  1-14-91  2:53p  ...A  Deleted:  3-08-91 11:08p
All of the clusters for this file are available. Undelete (Y/N)?y

File successfully undeleted.



                            Press any key to return to MS-DOS Shell
```

Fig. 5.1
Undeleting
files tracked
by the MIRROR
command.

5

*Just
Read*

Exercise 8.2: Undeleting a File

This exercise assumes that you have completed Exercise 8.1 in the same
computing session (you have not turned off your computer or rebooted since
doing Exercise 8.1). In this exercise, you delete a group of files, view a list of
the deleted files stored by MIRROR, and then undelete one of the files.

1. Set the default drive and directory to A:\TMP.
2. Delete all files beginning with the letter *D*. Use the wild card so that
   you can perform this operation with one command.
3. Use the /LIST switch of the UNDELETE command to list the files that
   have been deleted. Type

   undelete a:\tmp\d*.* /list

   Press ⏎Enter.
4. Now, undelete the files DOS.TXT and DOG.TXT. Again, use the wild
   card to do this in one command. Type

   undelete a:\tmp\do?.txt /dt

   Press ⏎Enter.
5. Follow the prompts, and type **Y** to undelete DOS.TXT and DOG.TXT.
6. View a listing of the files in the \TMP directory on drive A to make sure
   that DOS.TXT and DOG.TXT have been undeleted.

Using UNFORMAT To Recover a Formatted Disk

When you format a disk that contains data, DOS saves information about the files on the disk before DOS clears the file allocation table and directory. If you accidentally format a disk and do not write any new files to the disk, you can use UNFORMAT to recover the files of the disk.

When you format a disk, DOS saves the information needed to unformat the disk later, provided that you do not put any files on the disk.

The syntax for the UNFORMAT command follows:

> **UNFORMAT** *d:* /TEST/J/P/U/P

The switch /TEST simply shows you what the outcome of UNFORMAT would be, without actually executing the unformat. Using the /TEST switch first is always a good idea. To compare the MIRROR file with the disk and send that comparison to the printer, use the /J and /P switches. Actually to unformat the disk, simply use the UNFORMAT command and drive name. If you find that the MIRROR file is old or corrupt, use the /U and /P switches.

Make sure that you unformat a disk immediately after you accidentally format the disk and before you write any files on the disk. After you write files on a disk, the information that used to be on the disk is lost.

After the UNFORMAT command, you must confirm that you want to proceed, and then DOS recovers the files on the formatted disk. When FORMAT saves the UNFORMAT information, it runs MIRROR on the disk before it formats the disk. This process is called *safe formatting*. UNFORMAT then uses the MIRROR information to unformat the disk.

Exercise 8.3: Unformatting a Disk

In this exercise, you format a previously used floppy disk to simulate accidental formatting. Then you use the UNFORMAT command to retrieve files lost during formatting.

1. Boot your computer, if necessary. Insert the destination disk that you formatted and copied earlier, using the DISKCOPY command, into drive A.
2. Format the disk in drive A. Type **format a:**, and press ⏎Enter.
3. Follow the prompts to format the disk and add a volume label (see fig. 5.2).

```
SAMPLE   DOC     1427 02-14-91   3:24a
SAMPLE   TXT     3072 06-18-90  11:41a
       2 file(s)        4499 bytes
                     1209344 bytes free

C:\DATA\DOCS>format a:
Insert new diskette for drive A:
and press ENTER when ready...

Checking existing disk format
Saving UNFORMAT information
Verifying 1.2M
Format complete

Volume label (11 characters, ENTER for none)?

   1213952 bytes total disk space
   1213952 bytes available on disk

      512 bytes in each allocation unit
     2371 allocation units available on disk

Volume Serial Number is 3021-16F1

Format another (Y/N)?
```

Fig. 5.2
The prompts to
format a disk.

4. Now, you will unformat the disk. First, run an unformat test on the
   disk in drive A to see what will happen if you unformat. Type

   unformat a:/test

   Press ⏎Enter.

5. Look closely at the computer screen (as shown in figure 5.3), and
   determine what each segment of information means.

```
Searching disk for MIRROR image.

The last time the MIRROR or FORMAT command was used was at 15:41 on 03-10-91.

The MIRROR image file has been validated.

Are you sure you want to update the system area of your drive A (Y/N)? y

The system area of drive A has been rebuilt.

You may need to restart the system.

C:\DATA\DOCS>dir a:

 Volume in drive A has no label
 Volume Serial Number is 3739-16EF
 Directory of A:\

SAMPLE   DOC     1427 02-14-91   3:24a
SAMPLE   TXT     3072 06-18-90  11:41a
       2 file(s)        4499 bytes
                     1209344 bytes free

C:\DATA\DOCS>
```

Fig. 5.3
Unformatting
the disk.

Restoring from Backup Copies

Remember that you can take measures to prevent data loss, and in some cases,
you can recover from data loss, but these measures are not foolproof.

The only way to avoid data loss is to make sure that you always have a backup copy of every file.

The most important data-protection measure you can take is learning to make backup copies of all your disk files.

A good practice is to use DISKCOPY to copy every floppy disk that comes with any new software you buy, even before you install the software on your hard disk. Whenever you create or change a very important data file, immediately copy it to a floppy disk. You learned how to use COPY and DISKCOPY earlier in this chapter. To back up all your data for safekeeping, use BACKUP and RESTORE.

5

BACKUP and RESTORE have several advantages over COPY and DISKCOPY to back up all your data:

- You can back up an entire disk or directory structure with one command.
- You can back up files that are larger than the capacity of your floppy disks.
- You can choose to back up only those files that were created or changed since the last time you ran BACKUP.
- You need fewer floppy disks.

The rest of this chapter covers the BACKUP and RESTORE command techniques. With the examples in this chapter, you learn to back up and restore your entire hard disk or selected directories and files. You also learn the various switches to use to adapt BACKUP and RESTORE to your particular needs.

The BACKUP and RESTORE commands are effective insurance against file loss. You can protect against the loss of hours or weeks of work through methodical use of the BACKUP command to make backup disks of your files. Of course, you should also master the complement to BACKUP, the RESTORE command. RESTORE uses your backup disks to replace files lost from your hard disk. BACKUP is of no use if you don't know how to use RESTORE.

Many companies have devised stand-alone backup software packages. These packages are easier to learn than the DOS utilities. They also execute much faster and require fewer floppy disks. In time, you may purchase one of these packages and scoff at BACKUP and RESTORE. Nevertheless, until you do, you should understand these important DOS utilities.

In theory, all backup software serves the same purpose. If you understand the benefits and the dangers related to each command and each package, you can safely use these utilities.

If you have not yet tried to back up disk files, or if you are learning your way around DOS, the rest of this chapter is important. If you apply this information, you may never experience the shock that comes to those who lose files.

Preparing for the Backup

DOS or your computer may have several methods available to back up your files. Your computer, for example, may have a tape backup unit as part of its peripheral hardware. The methods for backing up files to tape also vary. You should know how to create and manipulate disk-based backups, however, in case you need to restore files to a computer that is not equipped with a tape backup.

With a single BACKUP command, you can back up an entire fixed disk. You can also use switches and parameters to make partial backups of the disk, backing up only selected files. You can select files by time, date, directory, activity, or file name.

Considering Full Backups

A *full backup* makes backup copies of all files on the hard disk, except for the hidden files, system files, and COMMAND.COM. A full backup transfers all other files on your hard disk to backup disks.

Deciding How Often To Back Up Hard Disks

Performing a complete backup about once a week is a good habit. If you do not regularly schedule partial backups to copy your most important files, do a complete backup more often.

On any day, ask yourself, "If my hard drive failed today, how much data would I lose?" Performing a backup is much easier than trying to reconstruct lost data. DOS does not prompt you to make backups. The decision is yours— remember, it's your data.

Preparing Backup Floppies

Before you perform a complete backup, make sure that you have enough floppy disks to hold all the files on your hard disk. You don't want to stop halfway through a backup to run to the computer store to buy more disks.

To determine how many megabytes of files are on your disk, use the CHKDSK command. The CHKDSK command is explained in more detail in Chapter 7. Divide the number of bytes by 1,000,000 to determine the number of megabytes you need to back up (see fig. 5.4).

5

Fig. 5.4
Calculating how
many floppy disks
you will need for
your backup.

```
C:\DATA\DOCS>chkdsk

Volume DRIVE_C     created 10-03-1990 3:45p
Volume Serial Number is 1650-7DB2

  27553792 bytes total disk space
     75776 bytes in 3 hidden files
    106496 bytes in 47 directories
  23617536 bytes in 975 user files
     20480 bytes in bad sectors
   3733504 bytes available on disk

      2048 bytes in each allocation unit
     13454 total allocation units on disk
      1823 available allocation units on disk

    655360 total bytes memory
    578064 bytes free

C:\DATA\DOCS>
```

After you know the approximate number of megabytes your files contain, you can use table 5.2 to estimate the number of disks you need. If you run out of disks, you can stop the backup by pressing [Ctrl] + [C].

Table 5.2 Floppy Disks Needed for Backup				
Megabytes	*Disk Capacity*			
	360K	*720K*	*1.2M*	*1.44M*
10M	29	15	9	8
20M	59	29	18	15
30M	83	44	27	22
40M	116	58	35	29
70M	200	100	60	50

One drawback to backing up a hard drive is the time it takes. But the hours, days, and often weeks of work you may one day save make that time well spent.

If you want, you can format the disks first. Do not use the /S switch when you format the disks; this switch decreases the available space on floppy disks because it copies system boot instructions during the format process. If you use unformatted disks, DOS formats each disk during the backup, but this option takes more time. Next, you should number each disk consecutively. BACKUP copies disks in sequential order so that RESTORE can put the files back on your hard disk in the proper order.

Although you can use older disks that contain files you don't want to keep, BACKUP overwrites old files only in the root directory. If you have an old disk with subdirectories, you should use FORMAT to delete all the files on the disk before you use the disk for backup.

Choose a regular interval to do a backup. Back up once each week or more often if you frequently change many files.

Get into the habit of performing a backup at least once a week, such as every Friday. Mark the dates on the calendar as a memory jogger. For very important files, use COPY or BACKUP to make a backup copy every time you change them.

Before you start the backup procedure, follow these steps:

1. Estimate how full your hard disk is.
2. Determine how many floppy disks you need for a backup (see table 5.2).
3. Make sure that you have enough floppy disks available to complete the backup.

The next section explains the actual backup procedure. Remember that BACKUP is half of a useful pair of commands. The other half, RESTORE, retrieves files copied with BACKUP.

Issuing the BACKUP and RESTORE Commands

The BACKUP command can copy files selectively from your hard disk to the destination floppy disk. The internal format of the backed-up file on the floppy is different from normal files. Therefore, you cannot use COPY to retrieve files stored on a backup disk. Your computer can use the files produced by BACKUP only after you run them through the RESTORE program.

Note: You should not attempt to back up or restore the hard drive on a computer that does not belong to you. If you're working on this text in a public computer lab at your school, read through this section. No exercises are given for this section, so simply familiarize yourself with the backup and restore procedures and outcomes described here.

Issuing the BACKUP Command

BACKUP can be a complex command to master because it has many options. You can execute BACKUP from the DOS Shell or the command line.

The syntax for the BACKUP command follows:

BACKUP *sd:spath\sfilename.ext dd: /S/F:size/M/A/D:mm-dd-yy*

where *sd:* is the letter of the drive containing the source disk (usually drive C), *spath* is the path to the files you want to make backup copies of, and *sfilename.ext* is the full file name of the file(s) you want to back up. Full file names may contain wild cards for selective backup of matching files. *dd:* refers to the drive that receives the backup files. Table 5.3 describes the optional switches that modify the basic BACKUP command.

Table 5.3	BACKUP Switches
Switch	*Description*
/S	Backs up subdirectories as well as the current directory
/F:*size*	Formats the target floppy disk to the size you specify (720K or 1.44M)
/M	Backs up files modified since the last backup. You use the /A switch with the /M switch to avoid erasing unmodified files when restoring from backup disks.
/A	Adds files to the files already on the backup disk.
/D:*mm-dd-yy*	Backs up files created or changed on or after the specified date

You must specify the same parameters whether you execute the program from the command line or from the Shell.

Performing a Full Backup

The full backup puts all files on your backup floppies. To do a full backup of drive C to drive A from the command line, enter the following command:

BACKUP C:*.* A: /S

This command tells DOS to back up all files in the root directory and to include all subdirectories (the /S switch). DOS prompts you to insert and change disks.

When you do a full backup from the DOS Shell, you can use the default parameters. Just choose OK from the dialog box (see fig. 5.5).

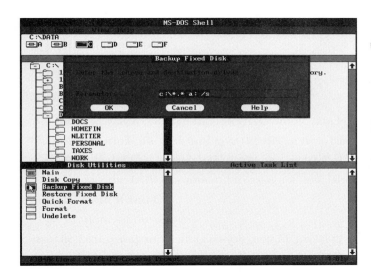

Fig. 5.5
Executing BACKUP
from the Shell.

Always write the backup date on the disks for future reference. Put the backup disks in the proper sequence, and store the disks in a safe place.

Performing Selective Backups

By specifying source directory paths, wild-card file names or extensions, and switches, you can select specific files to back up. Selective backups are useful when just some of your data changes between full backups, or if you move specific files from one computer system to another system.

Specifying Selected Directory and File Names

BACKUP always starts in the directory you specify in the source path. If you place a directory name in the path at a point in the tree other than the root directory, you can back up files in that directory (and its subdirectories with /S). To back up all files in the C:\DATA directory and all subdirectories, for example, enter the following command:

BACKUP C:\DATA*.* A:/S

Whether you execute BACKUP from the command line or the Shell, you see the same messages.

You can add further selectivity by using wild cards in the file name. To back up all files that have the extension DOC in the DATA directory and subdirectories, for example, enter the following command:

BACKUP C:\DATA*.DOC A:/S

The following section describes several other switches that are available.

Adding Other Switches

The /M (modified) switch selects only those files that have changed since the last backup. The /D and /T switches select files based on date and time.

You can enter a date and time after the appropriate switch, using the same formats you use for the DATE and TIME commands. You use these switches to include in the backup files changed on or after a specified date and time. You can use the /A switch to add files to a backup disk series and leave the existing backup files intact.

If the destination disk is not formatted, DOS formats the disk. If the disk does not match the maximum capacity of the drive, however, DOS formats the disk incorrectly. If you have a 1.44M floppy drive and you use an unformatted 720K disk, for example, DOS tries to format the disk as a 1.44M disk. To avoid this problem, use the /F switch and specify the size of the disk in the drive. To use an unformatted 720K disk in a 1.44M drive, for example, use the following parameter:

BACKUP C:\DATA*.DOC A: /F:360

The /F switch for formatting is explained in more detail in Chapter 3.

Issuing the RESTORE Command

The partner command to BACKUP is the DOS external command RESTORE. RESTORE is the only command that copies backed-up files to the hard disk. RESTORE's syntax is similar to BACKUP's syntax:

RESTORE *sd: dd:\dpath\dfilename.ext /switches*

where *sd:* refers to the source drive (the name of the drive holding the files you want to restore), *dd:* is the hard disk you want to restore to (usually drive C), and *dpath* represents the directory on the hard disk that receives the restored files. Files on the backup floppies that didn't come from the *dpath* directory are not restored. *dfilename.ext* is the file name for the file(s) to be restored. You can use wild cards in the file name to select specific files. */switches* stands for the optional switches you can add to gain further selectivity with the RESTORE command. Table 5.4 lists the switches available with the RESTORE command.

Table 5.4 RESTORE Switches	
Switch	*Description*
A:*date*	Restores all files created or changed on or after a specified date. This switch uses the DATE command format.
/B:*date*	Restores all files created or changed on or before a specified date. This switch uses the DATE command format.
/E:*time*	Restores all files modified at or earlier than the specified time. This switch uses the TIME command format.
/L:*time*	Restores all files modified at or later than the specified time.
/M	Performs similar to /N, but also restores files that have changed since the backup.
/N	Restores only files that are no longer on the hard disk. This switch is useful if you accidentally deleted hard disk files and need to restore them from backups.
/P	Prompts you to restore the file if it is marked as read-only or has been changed since the last backup.
/S	Restores subdirectories of the initial directory

Choose Restore Fixed Disk from the Disk Utilities Program Group to execute RESTORE from the Shell. The Parameters box defaults to a full restore of drive C from drive A. Choose OK to start the backup.

You must specify the same parameters whether you execute the program from the command line or from the Shell.

Performing a Full Restore

The full restore copies all the files on the backup disks to the hard disk. To use the command-line method to perform a full restore of drive C from drive A, enter the following command:

   RESTORE A: C:*.* /S

This command tells DOS to RESTORE all files in the root directory and to include all subdirectories (designated by the /S switch). DOS prompts you to insert and change disks.

When you do a full restore from the Shell, you can use the default parameters; just choose OK in the dialog box (see fig. 5.6).

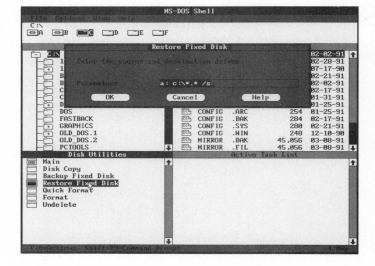

Fig. 5.6
Executing
RESTORE from
the Shell.

Performing Selective Restores

By specifying source directory paths, wild-card file names or extensions, and switches, you can select specific files to restore. Selective restores are useful when only one or a few files were destroyed or you want to restore the files to a different computer.

Restoring One File

You can choose to restore a single file by using a complete path and file name in the command. As in all selective restores, DOS prompts you to insert sequential disks until it locates the specified file.

Suppose that you want to restore the file \DATA\TAXES\CHECKS.WK1 from the backup disk. The proper command follows:

> **RESTORE A: C:\DATA\TAXES\CHECKS.WK1**

This command restores the CHECKS.WK1 file to its precise location.

Restoring Selected Directories and File Names

You can choose to restore more than one file. If, for example, you want to restore all files with a WK1 extension to the \DATA\TAXES directory, use the following command:

> **RESTORE A: C:\DATA\TAXES*.WK1**

The wild card *.WK1 selects all files with the WK1 extension from the DATA\TAXES directory. If you want to restore all files in a directory, and all subdirectories below the directory, use the /S switch, as in

> **RESTORE A: C:\DATA*.* /S**

This command restores all files in directories subordinate to the \DATA directory.

Note that when you use the RESTORE command in the same way as the BACKUP command, RESTORE acts on the same files.

5

Avoiding DOS Version Conflicts

Beginning with Version 3.3, DOS has a different method for producing the contents of a backup disk. DOS Versions 3.3 and later can restore files that you backed up with previous versions of DOS. Versions earlier than 3.3, however, cannot restore backups made with Versions 3.3, 4.*x*, or 5.

Although you cannot restore backed-up files from newer versions of DOS on computers running early versions of DOS, you can use COPY to move the files. Be careful when you use RESTORE or COPY to copy files from one computer to another computer running another version of DOS. Make sure that you do not copy any of the files in the /DOS directory. Each version of DOS has its own set of DOS files, and you cannot execute a command from one version of DOS on a computer running another version of DOS.

If you have a computer running MS-DOS 3.3, for example, and you copy the files from the /DOS directory of a computer that runs DOS 5, you get an error message every time you run an external command, such as CHKDSK, BACKUP, or RESTORE.

This problem occurs most often when you do a full backup on one computer and a full restore on another computer that runs a different version of DOS. The full RESTORE copies all the files except the hidden system files that make up the part of DOS you load into memory when you boot. This process results in a mismatch of DOS files, and you get errors when you try to run external DOS commands.

Chapter Summary

In this chapter, you have learned important concepts regarding file maintenance, management, and archiving practices. Although you will hear from every computer user how important it is to back up your files, most computer users learn this lesson the hard way. The underlying message of this chapter is to be careful with your data. You work many hours to fill floppy and hard disks with documents, and one careless command can wipe out those hours.

Testing Your Knowledge

True/False Questions

F 1. Relative-path commands are longer (require more keystrokes) than full-path commands.

T 2. It is always safer to use a full-path command.

True 3. You can copy *and* rename the destination file by using only one command.

T 4. The UNDELETE command enables you to undelete only files that have been captured by the MIRROR utility.

F 5. The BACKUP command is no more efficient or faster than the COPY command.

Multiple Choice Questions

1. To copy a file named WORDS.DOC from the \WP directory of drive C to the root directory of drive A (in the safest and most effective way), you should use the command

   A. COPY C:\WORDS.DOC A:

   B. COPY WORDS.DOC A:

   C. COPY WORDS.DOC A:WORDS.DOC

   D. COPY C:\WP\WORDS.DOC A:WORDS.DOC

   E. COPY A:WORDS.DOC C:\WP\WORDS.DOC

2. You want to use the best (shortest, most efficient) choice for copying all the following files to drive A: CHPT1.TXT, CHPT2.QUE, CHPT3.DOS, and CHPT4.MEM. Assuming that C:\WP is the default and that other files in the directory include CHPT10.TXT and CHP20.MEM, you should use the command

   A. COPY CHPT*.* A:

   B. COPY *?.* A:

   C. COPY CHPT?.* A:

   D. COPY C:\WP\CHPT*.* A:

   E. cannot do this in one command

3. To keep a log of all files deleted, you should use the command
    A. UNDELETE
    B. LOGDELETE
    C. UNFORMAT
    D. MIRROR
    E. COMPARE

4. The UNFORMAT command is effective
    A. when the disk has just been formatted and no new files have been copied on to the disk.
    B. when no more than five days have passed since the disk was formatted and used.
    C. when new files have been written to previously unused portions of the disk.
    D. only when MIRROR has been turned on.
    E. none of the above

5. To avoid data loss, you can
    A. back up files on a regular basis.
    B. use a surge protector on your computer.
    C. keep your computer area clean and smoke-free.
    D. use antistatic spray on the carpet around your computer area.
    E. all of the above

Fill-in-the-Blank Questions

1. Using the disk drive and directory name for both the source and destination files in the COPY command is called using the ___FULL path___.

2. The command to view deleted files stored by the MIRROR utility is ___UNDELETED D : patch\File.name.ext /List___

3. The command to view the effects of unformatting a disk without actually unformatting it is ___UNFORMAT D: TEST /J/P/U/P___

4. The command to make an identical duplicate of a floppy disk on to another floppy disk is ___UNFORMAT d: TEST /J/P/U/P. → Diskcopy___

5. The command to place files that have been saved with the BACKUP utility back on a disk is ___BACKUP C:\Data*.*.A:S___

Review: Short Projects

[handwritten: NANCEE OTT]

[handwritten left margin: Dr. Tim Gatenil / Kim / Pats Blackenship]

1. Learning How Others Protect against Data Loss

   Interview a computer consultant in a computer lab on your campus, and determine how the following issues are addressed:

   A. Unnecessary files on a hard drive *[handwritten: deleted]*

   B. Backups of software *[handwritten: Always have it]*

   C. Protection of software applications from illegal copying *[handwritten: NEVER COPY/ unless it FREE WARE/ soft.]*

   D. Protection from viruses *[handwritten: check Hard drive AND disk.]*

2. Protecting against Data Loss

   If you own your own computer, ask yourself the same questions you asked the computer consultant in Project 1. How do your answers compare?

[margin: 5]

3. Writing Commands for BACKUP and RESTORE Tasks

   Write out the necessary commands to complete the following tasks:

   A. Back up all files in the \BOOK subdirectory of drive C that have been changed since 4/1/93. *[handwritten: BACKUP c:\Book A:/D: 04-01-93]*

   B. Back up on to 720K floppy disks all files (including subdirectories) in the \BOOK subdirectory of drive C. *[handwritten: BACKUP c:\Book/s A: /F:720]*

   C. Restore from drive A to the \BOOK subdirectory of drive C all files that have been changed on or before 4/1/93. *[handwritten: RESTORE A: C:\Book /B : 04-01-93]*

Review: Long Projects

1. Backing Up the C:\DOS_WORK Directory

   For this project, you need an unformatted floppy disk. Run BACKUP on the C:\DOS_WORK directory. You should have your unformatted floppy disk in drive A before issuing the BACKUP command. Be sure to give the appropriate /F:*size* switch for your disk type. Compare the amount of file space required by the actual files in the C:\DOS_WORK directory to the file space taken up on the A: disk by the backed-up files.

2. Restoring the DOS_WORK Directory

   For this project, you restore to a subdirectory on your A: disk the files you backed up in Long Project 1.

Place into drive A the disk to which you backed up the files in Long Project 1. Make drive A the default drive. Create a directory on the A: disk called A:\DOS_WORK. Use the RESTORE command to restore the files in the root directory of drive A to the DOS_WORK directory in drive A. Here is the command you issue:

RESTORE A: A:\DOS_WORK

5

Beyond the Ten Commandments

6

In Chapter 3, you learned that every computer user needs to know about ten basic DOS commands. This chapter describes commands beyond the "basic ten." The purpose of this chapter is to familiarize you with some of the power that DOS provides so that when you require more than the basic ten, you will remember that these commands exist and look them up.

Objectives

1. To Determine the Amount of Memory Installed in Your Computer
2. To Examine the Content and Space Availability on a Disk
3. To Clear the Screen
4. To Determine the Version of DOS Installed on Your Computer
5. To Turn the VERIFY Option On and Off
6. To Determine the Volume Label of a Disk
7. To Change the Volume Label of a Disk
8. To Understand the DOS RECOVER Command
9. To Learn How To Redirect Output to Various Devices
10. To Use Pipes and Filters To Customize DOS Commands

Key Terms in This Chapter	
ASCII file	A file whose contents are alphanumeric and control characters; these characters can be text or other information you can read.
Lost clusters	Information on the hard disk lost to you because DOS lost the clusters' directory entries.
Fragmentation	One or more files that are physically spread over two or more separate parts of the disk. Fragmentation slows disk operations.
Noncontiguous	Means "not together." If a file is fragmented, it contains one or more noncontiguous blocks.
Defragment	To remove disk fragmentation so that every file on the disk is in one contiguous block.
Bad sector	A sector on a disk that contains a bad spot in the magnetic coating, where data cannot be read reliably.
Redirection	Receiving input from some place other than the keyboard or sending output to some place other than the screen.
Device	A hardware component or peripheral that DOS can use in some commands as though it were a file.
Console	The device DOS uses for keyboard input and screen output. DOS recognizes the console as CON.
Pipe	A method of taking the output of one command and making it the input for another command in DOS.
Filter	A method of processing the output of a command before displaying it.

6

Objective 1: To Determine the Amount of Memory Installed in Your Computer

The memory report utility tells you how much memory is installed on your computer, and how much is available at any given time. This command is especially useful if you are working on someone else's computer and don't know how much memory is available.

Following is the symbolic syntax for this command:

MEM */PROGRAM/DEBUG/CLASSIFY*

You can use the MEM command by itself to provide a current memory report. The /PROGRAM switch lists all programs loaded into memory and the amount of memory they are currently using. The /DEBUG switch displays program and internal computing commands for device drivers and the amount of memory they use, such as a video card or printer. The /CLASSIFY switch tells you whether programs are loaded into conventional memory (the first 512K of RAM) or upper memory (everything more than 512K).

6

Exercise 1.1: Determining the Amount of RAM in Your Computer

In this exercise, you use the MEM command to see how much memory is installed in your computer.

1. Boot the computer, if necessary.
2. At the DOS prompt, type **mem** and press ⏎Enter. Information similar to that in figure 6.1 appears.

How does the memory of your computer compare with the memory in the figure?

Objective 2: To Examine the Content and Space Availability on a Disk

CHKDSK is the DOS command that checks disk space and provides a detailed report of disk and memory status. CHKDSK also can repair certain errors on the disk.

```
655360 bytes total conventional memory
655360 bytes available to MS-DOS
535392 largest executable program size

983040 bytes total EMS memory
491520 bytes free EMS memory

2490368 bytes total contiguous extended memory
 983040 bytes available contiguous extended memory
      0 bytes available XMS memory
        MS-DOS resident in High Memory Area



                    Press any key to return to MS-DOS Shell
```

Fig. 6.1
The amount
of RAM in a
computer.

Following is the symbolic syntax for the checkdisk command:

CHKDSK *d:path.ext* /F/V

The /F switch fixes problems the utility finds on the disk. The /V switch is used for *verbose display mode*, which causes the CHKDSK utility to display all directories and files as it checks them. This display can amount to many screens of information if you are searching a large hard drive.

Before CHKDSK displays its status report, it checks every file on the disk and makes sure that the information in the file allocation table (FAT) agrees with the information in the disk directory. Any errors represent *lost clusters*, or files that are lost. You can lose files if a program freezes as it updates a file or if you turn off or boot the computer as a program updates a file.

If CHKDSK finds any errors, it displays a warning message. You must use the /F switch to fix these errors.

Exercise 2.1: Examining Your Disk with CHKDSK

In this exercise, you view the hard drive with the CHKDSK command.

1. Boot your computer, if necessary.
2. At the C: DOS prompt, type **chkdsk** and press ⏎Enter . A report similar to the report in figure 6.2 appears.

Compare the CHKDSK report your computer generated with the one in figure 6.2.

```
C:\DATA>chkdsk

Volume DRIVE_C    created 10-03-1990 3:45p
Volume Serial Number is 1650-7DB2

  27553792 bytes total disk space
     75776 bytes in 3 hidden files
    108544 bytes in 47 directories
  23654400 bytes in 989 user files
     20480 bytes in bad sectors
   3694592 bytes available on disk

      2048 bytes in each allocation unit
     13454 total allocation units on disk
      1804 available allocation units on disk

    655360 total bytes memory
    291152 bytes free

C:\DATA>
```

Fig. 6.2
A status report
of the disk and
available memory.

6

Using CHKDSK /F

Proper use of the /F switch is a common "fix" to convert lost clusters on disks into files. The /F switch sometimes can recover important data you may have lost.

The /F switch causes DOS to prompt you as to whether it should proceed before fixing problems detected by the command (see fig. 6.3).

```
C:\>CHKDSK /F

Volume BILLS ACER  created 07-24-1990 8:10p
Volume Serial Number is 3E40-1BCF

       138 lost allocation units found in 2 chains
Convert lost chains to files (Y/N)?Y

  97728512 bytes total disk space
    120832 bytes in 7 hidden files
    143360 bytes in 55 directories
  53618688 bytes in 2116 user files
    282624 bytes in 2 recovered files
  43563008 bytes available on disk

      2048 bytes in each allocation unit
     47719 total allocation units on disk
     21271 available allocation units on disk

    655360 total bytes memory
    389088 bytes free

C:\>
```

Fig. 6.3
Result of the
/F switch.

You can safely delete files that are no longer useful. CHKDSK converts clusters into files with the name FILEnnnn, in which FILEnnnn stands for FILE0001, FILE0002, FILE0003, and so on (see fig. 6.4).

```
C:\>DIR *.CHK

 Volume in drive C is BILLS ACER
 Volume Serial Number is 3E40-1BCF
 Directory of  C:\

FILE0000 CHK      4096 09-12-90  12:06p
FILE0001 CHK    278528 09-12-90  12:06p
        2 File(s)   43563008 bytes free

C:\>
```

Fig. 6.4
CHKDSK converts
clusters into files.

Avoid using /F until you know the implications of the fix action. For more information about the /F switch, refer to Que's *MS-DOS 5 User's Guide*, Special Edition.

Caution: Do not run CHKDSK /F from the DOS Shell or when any other program is running. CHKDSK may consider a file in use by the other program to be a lost cluster, and you could lose part of a good file.

Exercise 2.2: Using CHKDSK /F/V To Examine a Floppy Disk

The checkdisk utility is a valuable tool for floppy disks as well as hard disks. For this exercise, you need a formatted floppy disk (preferably one that contains files).

1. Boot the computer (if necessary), and insert a formatted floppy disk into the A drive.
2. Change the default drive to A.
3. Type **chkdsk a:** /f/v at the A: prompt, and press ⏎Enter .

Did the CHKDSK utility find any problems?

Using CHKDSK before a Backup

One useful feature of CHKDSK is its capability to report the number of bytes used in directories, user files, and hidden files. You can use CHKDSK before you issue the BACKUP command to determine the number of bytes to be backed up. You then can determine the number of floppy disks you need for the backup.

152 DOS

Space on floppy disks usually is measured in kilobytes or megabytes. Divide the total number of bytes to be backed up by 1,000 for kilobytes, or 1,000,000 for megabytes. To determine how many disks you need to complete the backup, divide the number of kilobytes or megabytes by the capacity of your backup floppy disks.

Suppose that you want to back up your C hard disk drive; however, you don't know how many floppy disks you need. To find out, issue the **CHKDSK C:** command.

The report gives you the disk's total capacity in bytes, the number of hidden files and their byte total, the number of directories and their byte total, and the number of user files and their byte total. The report also gives the total memory bytes in RAM and the number of RAM bytes free for use.

The total number of bytes in directories and user files is the number of bytes to be backed up. The directory files normally do not add up to much more than 100K. You will be safe using the user files total to calculate the number of floppy disks you need for a backup.

6

Note: Remember that the number of disks required is based on their capacity to hold a specific amount of data. A 1.2M floppy disk, for example, holds four times the data of a 360K disk. Verify that your disks have the capacity you expect. Double-check disk labels for the correct density, and watch for bad sectors when formatting floppy disks.

Understanding CHKDSK and Fragmentation

As you add and delete files on a disk, the free space for new files spreads around the surface of the disk.

When you create or change a file and your application program or a DOS command writes a file to disk, DOS allocates data storage space by finding the next available disk space. If the first available space is too small to hold the entire file, DOS fills that space and places the remainder of the file into the next available space(s).

This phenomenon is called *fragmentation*. Even technological marvels can be sloppy housekeepers; fragmented files reduce disk performance.

Following is a form of the CHKDSK command that checks for fragmentation:

CHKDSK *d:filespec*

d: is the drive letter of the disk you want to check. *filespec* is the file specification for the file or files you want to check for fragmentation. When a file is fragmented, CHKDSK reports the number of *noncontiguous* blocks used to contain the file. Each block is a separate area of the disk used to hold part of the file.

CHKDSK reports whether a file is fragmented (not stored in a contiguous block). In figure 6.5, all files in the current directory are checked, and CHECKS.WK1 is fragmented.

Fig. 6.5
A CHKDSK report.

```
C:\DATA\TAXES>CHKDSK *.*

Volume DRIVE_C     created 10-03-1990 3:45p
Volume Serial Number is 1650-7DB2

  27553792 bytes total disk space
     75776 bytes in 3 hidden files
    108544 bytes in 47 directories
  23652352 bytes in 991 user files
     20480 bytes in bad sectors
   3696640 bytes available on disk

      2048 bytes in each allocation unit
     13454 total allocation units on disk
      1805 available allocation units on disk

    655360 total bytes memory
    324880 bytes free

C:\DATA\TAXES\CHECKS.WK1 Contains 7 non-contiguous blocks

C:\DATA\TAXES>
```

Fragmentation is not serious—it happens to all active disks; however, fragmentation slows disk operations. You can defragment a disk in two ways. The most common and the fastest is to buy a DOS utility package, such as PC Tools Deluxe, Mace Utilities, or Norton Utilities, that contains disk defragmentation programs.

Another way to defragment a disk is to back up all the files to another disk, delete the files from the hard disk, and then restore the files to the hard disk. For more information about the BACKUP and RESTORE commands, see Chapter 5, "Maintaining Files."

Objective 3: To Clear the Screen

The CLS command erases, or clears, the display and positions the cursor at the top of the screen, after the DOS prompt. Use CLS when the screen becomes too "busy" with the contents of previous commands' output. CLS has no parameters; thus, the symbolic syntax is

CLS

You simply type **cls** at the DOS prompt, and the screen clears. The CLS utility frequently is used in batch files that display on-screen messages to the user.

Exercise 3.1: Clearing the Screen of the Computer

To practice clearing the computer screen, follow the steps in this exercise:

1. Boot the computer, if necessary.
2. Perform a DIR of the default disk so that information appears on-screen.
3. Type **cls** at the DOS prompt, and press ⏎Enter.

The screen clears immediately, and the cursor reappears at the top of the screen.

Objective 4: To Determine the Version of DOS Installed on Your Computer

You may find it useful to know the exact version of DOS your computer is using. The VER command reports a manufacturer's name and the version number of DOS. Because of the version of DOS you are using, your particular VER report may look slightly different from that in figure 6.6.

```
C:\>ver

MS-DOS Version 5.00 RC 3a

C:\>
```

Fig. 6.6
Using the VER command.

VER is useful if you must work on another person's computer. To find out which DOS version the computer is using, issue this command before you start work. Then you know which commands or switches the computer accepts.

VER is also useful when you boot your system from a disk you did not pre-pare. The floppy disk may contain system files from a version of DOS you do not use normally. The VER command has no parameters and is issued at the DOS prompt; thus, the symbolic syntax is

> VER

Exercise 4.1: Checking Your DOS Version

In this exercise, you determine which version of DOS is installed on your computer.

1. Boot the computer, if necessary.
2. Type **ver** at the DOS prompt, and press ↵Enter.

Which version of DOS are you using?

Objective 5: To Turn the VERIFY Option On and Off

VERIFY checks the accuracy of data written to disks during copy or move (in DOS 6) operations. VERIFY has one parameter, which is ON or OFF. The symbolic syntax is:

> **VERIFY ON|OFF**

When you type **verify on** at the prompt, DOS rereads all data to ensure that it was written correctly. When VERIFY is set to ON, DOS operations are slower; therefore, you may want to use the command with only important data. Type **verify off** to turn off the verification.

Exercise 5.1: Verifying a Copied File

You need a formatted floppy disk for this exercise.

1. Boot the computer, if necessary, and insert a formatted floppy disk into the A drive.
2. To turn on the verify option, type **verify on** and press ↵Enter.
3. To check the status of verify to see that it is on, type **verify** and press ↵Enter. DOS confirms that VERIFY is on (see fig. 6.7).

```
C:\>verify
VERIFY is on
C:\>
```

Fig. 6.7
Verifying a
copied file.

6

4. To copy the file MISER.TXT from the DOS_WORK subdirectory
    of the hard drive to your floppy disk in the A drive, type
    copy c:\dos_work\miser.txt a: and press ⏎Enter . Notice how
    long DOS takes to copy this very small file.
5. To turn off verify, type **verify off** and press ⏎Enter .

Objective 6: To Determine the Volume Label of a Disk

When you format a disk, you can enter an internal disk volume label for the
disk. Following is the symbolic syntax for this utility:

   VOL *d*:

When you type **vol** at the DOS prompt, if you entered a name when you
formatted the disk, DOS displays the volume name of the current disk (see
fig. 6.8).

The VOL command also displays the volume serial number assigned automatically by DOS when you formatted the disk. If you add a drive letter after the
command, DOS displays volume information for that drive.

Viewing volume names is much easier than wading through directory listings
when you sort through floppy disks.

```
C:\>VOL

 Volume in drive C is DRIVE_C
 Volume Serial Number is 1650-7DB2

C:\>VOL B:

 Volume in drive B has no label
 Volume Serial Number is 1935-13E7

C:\>
```

Fig. 6.8
The VOL
command.

Exercise 6.1: Checking the Volume Label

You need a formatted floppy disk for this exercise.

1. Boot the computer (if necessary), and insert a formatted floppy disk into the A drive.
2. To check the volume label of the hard drive, type vol c: and press ⏎Enter.
3. Check the volume label of the floppy disk in the A drive.

Objective 7: To Change the Volume Label of a Disk

After you format a disk, you can use the LABEL command to add or change the volume label. When you change the use for the disk, you should change the volume label of the disk. When you first formatted the disk, for example, you may have labelled it TAX92. To use the disk to keep a backup copy of your 1993 business files later, change the label to BUSINESS93.

Following is the syntax for this command:

> **LABEL** *d: label*

d: is the drive letter, and *label* is the new label. If you omit the drive letter, DOS uses the current drive. If you omit the label, DOS prompts you to enter the label, or you can press ⏎Enter for no label.

When you do not specify a label, DOS displays the current volume information and prompts you for a label. To remove an existing label, just press ⏎Enter at the prompt (see fig. 6.9).

```
C:\>LABEL B:
Volume in drive B is TAX91
Volume Serial Number is 1935-13E7
Volume label (11 characters, ENTER for none)? BUSINESS92

C:\>
```

Fig. 6.9
The LABEL
command.

Exercise 7.1: Changing the Volume Label of Your Floppy Disk

You need a formatted floppy disk for this exercise.

1. Boot the computer (if necessary), and insert a formatted floppy disk into the A drive.

2. Because you're using this disk as a generic work disk, name the disk WORK. To do this, type label a: work and press ⏎Enter.

Objective 8: To Understand the DOS RECOVER Command

When you save a file in an applications program or copy a file with DOS, you expect that file always to be available when you need it. You learned in Chapter 5, however, that files can become corrupted. Because there are many threats to your data, you must always have a backup copy of every important file.

One threat to your data is a bad spot on the disk itself. Data is magnetically recorded on a very thin coat of iron oxide. The high-speed manufacturing process that coats the platters of a hard disk is not perfect. Many hard disks have a few spots where the oxide coat is too thin or imperfect and cannot record data reliably.

When you format a disk, DOS checks the disk surface and marks as bad sectors any sectors that it cannot read. After a sector is marked as being bad, DOS ensures that data is never recorded on that sector. You can find out the number of bad sectors on a disk when you format the disk or when you execute the CHKDSK command.

A few bad sectors on a hard disk are not a problem. A floppy disk with bad sectors, however, may indicate a problem with the entire disk. Do not store important data on a floppy disk that contains bad sectors. Figure 6.10 shows a report in which the bytes in bad sectors represent less than 0.1 percent of the total capacity of the disk.

Fig. 6.10
Running CHKDSK
enables you to see
information about
a disk.

```
C:\>CHKDSK

Volume DRIVE_C      created 10-03-1990 3:45p
Volume Serial Number is 166D-BDE3

  27553792 bytes total disk space
     75776 bytes in 3 hidden files
    112640 bytes in 48 directories
  24518656 bytes in 916 user files
     20480 bytes in bad sectors
   2826240 bytes available on disk

      2048 bytes in each allocation unit
     13454 total allocation units on disk
      1380 available allocation units on disk

    655360 total bytes memory
    592768 bytes free

C:\>
```

DOS takes care of bad sectors automatically. Unfortunately, a spot on the disk may be marginal or just good enough to pass the bad-sector test when you format the disk. As you read and write files, DOS no longer may be able to read data from this spot: a marginal sector has become a bad sector.

When DOS tries to read data from a bad sector, it returns an error. The error message usually is sector not found or seek error. The worst time to get this message is when you try to make a backup copy of the file.

If the unreadable file contains mostly text, such as a word processing file, you may be able to recover most of the data. The RECOVER command reads the file one sector at a time and restores the part of the file that can be read. The part of the file in the bad sector is lost. The format is as follows:

> **RECOVER** *filespec*

The *filespec* in most cases is the name of the file you want to recover. Although *filespec* may be a directory or even an entire disk, you should recover only one file at a time. No matter where the file started originally, RECOVER places the recovered file in the root directory and assigns the name FILE*nnnn*.REC. The *nnnn* is a sequential number beginning with 0001. If you recover one file at a time, you then can rename the file to its correct name. If you try to recover multiple files simultaneously, you will need to read through each FILE*nnnn*.REC to determine particular files.

After you recover a file, use a word processing program or the DOS Editor to determine the missing data, and then reenter that data. Because of the missing data, you may get an error when you try to read the file into a word processing program, you will see a message such as Incomplete file or Invalid file format. If the program refuses to accept the file, use the DOS Editor to delete all the non-text formatting information and read the file into your word processing program as a text file. You will need to reenter all format information.

If the recovered file is a database or a spreadsheet, you may be able to use the recovered file, or depending on the part of the file that was lost, the program may just reject it. If the recovered file is a program, the recovered file is useless. Do not try to rename and execute a recovered program file. If you try to execute an incomplete program, the results are unpredictable and could be disastrous.

DOS 6.0 Note: The RECOVER command for DOS Version 6.0 is phased out; however, many commercial utilities that do a much better job of file recovery than the DOS 5.0 recover utility are available.

6

Objective 9: To Learn How To Redirect Output to Various Devices

The term *redirection* in DOS means to change the source or destination normally used for input and output. The standard source of input is the keyboard. The standard output location is the screen display. When you use the keyboard to type a command, COMMAND.COM carries the text or messages and displays them on-screen. The keyboard and display in DOS are the standard, or default, devices for messages, prompts, and input.

DOS views devices as extensions of the main unit of the computer. Some devices send input and receive output. Other devices are used for input only (the keyboard) or output only (the video display or printer) (see fig. 6.11).

Disk drives are both input and output devices. Keyboards are input devices; displays and printer adapters are output devices. Serial adapters can send output and receive input.

You can use device names, as you do file names, in some commands (see table 6.1). In fact, DOS treats devices as if they were files. Device names are three or four characters long and do not contain extensions. You cannot delete device names or format them, but you can use device names in commands to perform some useful actions.

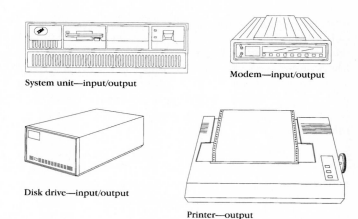

System unit—input/output

Modem—input/output

Disk drive—input/output

Printer—output

Fig. 6.11
Input and output
sources.

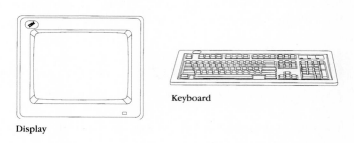

Display

Keyboard

DOS controls devices through its system files and the ROM BIOS. Fortunately, understanding the details of how DOS handles devices is not essential to using them with DOS commands.

COPY CON

A useful application for a DOS device is creating a file containing characters you enter directly from the keyboard. To do this, you use the familiar COPY command. This time, however, you copy data from CON, the *CONsole* device.

Following is the symbolic syntax for COPY CON:

  COPY CON *d:path*

You use this command frequently when you create batch files (see Chapter 7).

Table 6.1 DOS Device Names	
Name	*Device*
CON	Video display and keyboard
AUX or COM1	First asynchronous communications port
LPT1 or PRN	First line or parallel printer. This device is used only for output.
LPT2	Second parallel printer
LPT3	Third parallel printer
COM2	Second asynchronous port
COM3	Third asynchronous port
COM4	Fourth asynchronous port
NUL	Dummy device for redirecting output to "nowhere"

Exercise 9.1: Creating a Small Text File

You need a formatted floppy disk for this exercise.

1. Boot the computer (if necessary), and insert the formatted floppy disk into the A drive.
2. Enter your name, address, city, state, and ZIP code into a file called AD.TXT, as in the following example:

   Type **copy con a:ad.txt**, and press ⏎Enter.

   Type **Elizabeth Samuels**, and press ⏎Enter.

   Type **4219 Kensington Place**, and press ⏎Enter.

   Type **Carmel, IN 46032**, and press ⏎Enter.
3. To save the file, press F6.
4. To view the contents of the AD.TXT file, type **type A:ad.txt** and press ⏎Enter.

Redirection

To instruct DOS to use nondefault devices in a command, you must use special redirection symbols. Table 6.2 lists the symbols DOS recognizes for redirection.

Table 6.2 Symbols for Redirection	
Symbol	*Redirection*
<	Redirects the input of a program
>	Redirects the output of a program
»	Redirects the output of a program and adds the text to an established file, if a previously created file exists

The < (less than) symbol points away from the device or file and says, "Take input from here." The > (greater than) symbol points toward the device or file and says, "Put output there." The » (chevron) symbol redirects a program's output but adds the text to an established file. When you issue a redirection command, place the redirection symbol after the DOS command but before the device name.

Exercise 9.2: Redirecting to a Printer

Wouldn't it be practical to get through your printer the output of a DIR or TREE command? You can by redirecting output to the device PRN (the printer). You need a printer for this exercise.

1. Boot your computer, if necessary.
2. Turn on your printer, and be sure that it is on-line.
3. To print a directory in five-column format, type **dir /w > prn** and press ⏎Enter.

When you press ⏎Enter, the output of the DIR command goes to the printer. The /W (wide display) switch lists the files in five columns. You can tuck such a printout into the sleeve of a floppy disk to identify the contents of a disk.

Exercise 9.3: Redirecting a Text File to the Printer

For this exercise you print the AD.TXT file you created in Exercise 9.1.

1. Boot your computer (if necessary), and insert into the A drive the floppy disk that contains the AD.TXT file.
2. Turn on your printer, and be sure that it is on-line.
3. To redirect the output of AD.TXT to the printer, type
   type ad.txt > prn.

You now have a printed copy of the AD.TXT file.

Caution: Never try to redirect binary files to the printer. Redirecting binary data can result in paper-feed problems, beeps, meaningless graphics characters, and maybe a locked computer. If you get hung up, you can do a warm boot or turn the power switch off and then on again. If you must turn your computer off, it is good practice to wait about 15 seconds before turning it back on.

Objective 10: To Use Pipes and Filters To Customize DOS Commands

The output of one command can be presented as the input of another command. The pipe symbol (|) presents (or *pipes*) output from one command to another. (Be careful that you don't confuse the pipe symbol with the colon symbol. The pipe symbol is on the same key as the backslash. Books often print the pipe symbol as a single vertical line but on the keyboard and on your display, it is two stacked vertical lines.) The FIND, SORT, and MORE commands are filters. A filter is a program that accepts data from the standard input, changes the data, and then writes the modified data to the display.

FIND, SORT, and MORE accept the output of a DOS command and perform further processing on that output. FIND outputs lines that match characters given on the command line. SORT alphabetizes output. MORE displays a prompt when each screen of the output is full.

Exercise 10.1: Using the FIND Filter

[handwritten: COPY FILES FROM G: C:?]

One handy way of using the FIND filter is to request a directory list that displays only the files that contain certain characters in its file name.

1. Boot your computer, if necessary.
2. Change directories to the DOS_WORK subdirectory on the C drive.
3. To list the files that contain *TAX* in the file names, type **dir | find "tax"** *[handwritten: caphism]* and press ⏎Enter.

The FIND command filters the output of a DIR command. FIND displays on-screen only the lines that contain TAX. The | symbol pipes the output of DIR to FIND (see fig. 6.12).

Fig. 6.12
The FIND command filters output of a DIR command.

```
Volume Serial Number is 166D-BDE3
Directory of C:\DATA\TAXES

.              <DIR>        01-28-91   3:18p
..             <DIR>        01-28-91   3:18p
CHECKS   WK1   37365 01-16-91   1:14p
INCOME   WK1    3622 09-18-90   1:50p
LOG      DOC    3072 06-18-90  11:41a
SAMPLE   WK1    1427 02-14-91   3:24a
STATEMNT WK1    3304 06-25-90   4:21p
TAXEST   WK1    5379 01-14-91   2:53p
1991TAX  DOC     223 03-15-91   3:56p
92TAX    WK1     252 03-15-91   3:57p
STATE    TAX     440 03-15-91   3:58p
       11 file(s)       55164 bytes
                      2824192 bytes free

C:\DATA\TAXES>DIR | FIND "TAX"
 Directory of C:\DATA\TAXES
TAXEST   WK1    5379 01-14-91   2:53p
1991TAX  DOC     223 03-15-91   3:56p
92TAX    WK1     252 03-15-91   3:57p
STATE    TAX     440 03-15-91   3:58p

C:\DATA\TAXES>
```

Exercise 10.2: Using the SORT Filter

The SORT command is a filter that alphabetizes input. The main use for the SORT filter with earlier versions of DOS was to sort in alphabetical order a directory listing. With DOS 5, to sort a directory listing by file name, use the /ON switch. You practice using the /ON switch in this section, as well as sorting names with the SORT filter in a text file.

1. Boot your computer, if necessary.
2. Change directories to the DOS_WORK subdirectory of the C drive.
3. To list the contents of the directory in alphabetical order, type **dir /on** and press ⏎Enter.

4. Now you use the SORT filter to sort a list of names within a text file. The output of the TYPE command in the following command is piped to SORT. Type **type names.txt | sort**.

When the listing filtered by SORT appears on-screen, the list is in alphabetical order (see fig. 6.13).

```
C:\DATA\DOCS>TYPE NAMES.TXT
Mallory
Lee
Wouk
Laun
Anderson
Kincaid
Shaw
Hofstadter
Andersen

C:\DATA\DOCS>TYPE NAMES.TXT | SORT
Andersen
Anderson
Hofstadter
Kincaid
Laun
Lee
Mallory
Shaw
Wouk

C:\DATA\DOCS>
```

Fig. 6.13
The contents of a directory sorted in alphabetical order.

6

Exercise 10.3: Using the MORE Filter

If a file is so long that text scrolls off-screen, you can pause the display by pressing Ctrl+S (or on enhanced keyboards, press Pause). If you have a fast computer, pressing Pause at the right time may be difficult, and information may scroll off-screen before you can read it. The MORE filter provides a better solution to this problem. MORE displays information one screen at a time.

1. Boot your computer, if necessary.
2. Change directories to the DOS_WORK subdirectory.
3. To view the way in which MORE works, type **type readme.txt | more** and press ⏎Enter.

MORE displays 23 lines of text and pauses while displaying the message — More —. When you press any key, MORE displays the next screen of text (see fig. 6.14).

```
README.TXT

NOTES ON MS-DOS VERSION 5.0
===============================
This document provides important information that is not included
in the Microsoft MS-DOS User's Guide and Reference or in online
Help. The following topics are covered:

1. How MS-DOS 5.0 Modifies Your System
     1.1 LINK.EXE Versions May Be Left in Path
     1.2 WINA20.386 File
2. Solving Setup Problems
     2.1 Priam and Everex Hard Disks
     2.2 Syquest Removable Hard Disk
     2.3 Bernoulli Drives
     2.4 Disk Manager
     2.5 Speedstor and Volume Expansion
     2.6 Novell Partitions
     2.7 Vfeature Deluxe
     2.8 Difficulty Specifying a DOS Path
     2.9 Incompatible Partition
-- More --
```

Fig. 6.14
The MORE filter.

6

Chapter Summary

In this chapter, you learn and use many useful DOS commands. The point of introducing these more advanced commands to you is to acquaint you with their utility. You probably will not remember these commands verbatim; however, you now know that they exist. When you must use one of the commands, you will know that DOS has the capability, and you can look in this text (or a DOS manual) to find them.

Testing Your Knowledge

True/False Questions

1. The DOS MEM command clears all unnecessary data from RAM.

2. CHKDSK /F is used to find files on a large hard drive.

3. The DOS RECOVER command is used to piece together damaged files, bit by bit.

4. The only way to get a printout of a directory listing in DOS is to display the directory on the screen, and then press PrtSc.

5. The vertical bar, called a pipe, is used to take the output of one DOS command and use that output as input for another DOS command.

4, 5, 6, 7

ASCII
ACII

AS1
META STREM
TE Estausture
redirection
Simbles

purpose of
simples.

Multiple Choice Questions

1. The best way to determine the number of disks needed for a backup procedure is
   - A. to perform a DIR of the hard drive.
   - B. to perform a MEM of the hard drive.
   - C. to use VER to determine the DOS version.
   - D. to VERIFY the hard drive.
   - E. to use CHKDSK on the hard drive.

2. The computing problem that occurs when only small space locations are available on a disk is called
   - A. compression.
   - B. fragmentation.
   - C. minimizing.
   - D. a bottleneck.
   - E. debugging.

3. The best way to change the internal volume label of a disk is to
   - A. reformat the disk and give the disk a new label when prompted.
   - B. use the VOL command to assign a new label.
   - C. use the LABEL command to assign a new label.
   - D. both B and C
   - E. none of the above

4. The fastest way to create a simple text file in DOS is to
   - A. use the TYPE command.
   - B. use a word processing program.
   - C. use COPY CON.
   - D. All of the above are equally fast.
   - E. none of the above

5. Which of the following commands displays a directory listing one screen at a time?
   - A. DIR /W
   - B. DIR /P
   - C. DIR |MORE
   - D. DIR |PAGE
   - E. none of the above

6

Fill-In-the-Blank Questions

1. The command to clear the computer screen is ___*cls*___.
2. The command to double-check that a file has been accurately copied is ___*VERIFY*___.
3. The symbol to redirect output is ___*> PRN*___.
4. MORE, SORT, and FIND are called ___*pipes and Filter commands*___
5. With a pipe, you redirect the ___*command INPUT*___ of one DOS *vertical bar* command so that it is the _____ for another DOS command.

Review: Short Projects

1. **Running CHKDSK**

   Run a CHKDSK on your floppy disk, and redirect the output to the printer. Turn the printout in to your instructor.

2. **Checking for Fragmentation** *not do*

   Choose three files located on your hard drive, and run CHKDSK on each file, redirecting the output of all three to the printer. Determine, from the printout, if fragmentation is a problem on your particular hard disk.

3. **Using FIND** *copy student files.*

   Using the FIND filter, compile a listing of all files that end in .BAT. Redirect this listing to the printer, and turn the printout in to your instructor.

Review: Long Projects

1. **Using COPY CON**

   Using the COPY CON utility, create a text file called BIRTHDAY.TXT, that contains the names and birthdates of six of your friends. The format for each line should be as follows:

   lastname, firstname, mm/dd/yy

When you finish creating the file, direct the contents of the file to the printer. In a second printout, sort the contents of the file before redirecting them to the printer.

2. Using the TREE Command

   Using the TREE command with redirection, print a copy of the structure of your hard drive. Compare this printout with the hand-written tree drawing you created in Long Project 2 in Chapter 2.

CD\ TREE > PRN

6

Customizing DOS

7

In this chapter, you create and use batch files. A *batch file* contains one or more DOS commands to be executed (carried out) as a group at the same time. Batch files enable you to perform redundant sequences of DOS commands with just a single command. For this reason alone, batch files are useful timesaving and effort-saving tools.

This chapter is not designed to make you a batch file expert, although you may be surprised to find how much you can learn in a few pages.

If you have read this far, you are a reasonably skilled DOS user. The following sections teach you how to create and modify the AUTOEXEC.BAT and CONFIG.SYS batch files. Also included in this chapter are some useful sample batch files.

Objectives

1. To Use the PATH Command
2. To Change the DOS Prompt by Using the PROMPT Command
3. To Understand Batch Files
4. To Create a Batch File
5. To Understand the AUTOEXEC.BAT File
6. To Create an AUTOEXEC.BAT File
7. To Understand the CONFIG.SYS File

Key Terms in This Chapter	
PATH command	The command that instructs DOS to search through a specified set of directories for programs or batch files. If you try to execute a program or batch file that is not in the current directory, DOS searches through the directories in the search path. If the program or batch file is found in one of the directories in the search path, DOS executes the program or batch file.
Batch file	A text file that contains DOS commands, which DOS executes as though the commands were entered at the DOS prompt. Batch files always have a BAT extension.
Metastring	A series of characters that to DOS takes on a different meaning from the literal meaning. DOS displays substitute text when it finds metastrings in the PROMPT command. The wild-card characters (* and ?), which you used in previous chapters, are one type of metastring.
AUTOEXEC.BAT file	A batch file that executes automatically each time you boot your computer. AUTOEXEC.BAT is an ideal place to include commands that initialize and personalize the control of a PC.
CONFIG.SYS file	A file used by DOS to tailor hardware devices and to assign the computer's resources.
Buffer	An area of RAM allocated by DOS as a temporary storage area for data that is moving between the disks and an executing program.

Objective 1: To Use the PATH Command

Using the PATH command is one of the most important techniques you can employ to make your computer a little easier to use.

When you tell DOS to execute an internal command that is built into COMMAND.COM (such as DIR), DOS has that command in memory and ready to execute. When you execute an external command (such as FORMAT), DOS must find the FORMAT program on your disk. First, DOS looks in the current directory. If DOS cannot find the program, DOS looks in the directories specified in the PATH command.

When the DOS installation set up a PATH command that included C:\DOS, the program made sure that you can use FORMAT and all the other external DOS commands no matter what the current directory is. You can add other directories to the PATH command so that you can execute batch files and other programs from any disk or directory on your computer.

Actually, DOS can find programs and batch files in these three situations:

* The file is in the directory in which you are working (the current directory). This situation is the case when you execute 1-2-3 from the directory that contains the 1-2-3 program files, such as 123R22, as explained in Chapter 4.
* The file is not in the current directory, but you include the full path on the command line.
* The directory in which the file is located is on the search path established by the PATH command. This situation is the case when you execute external DOS commands and, as you will see, when you execute batch files.

The search path is the list of directories specified in the PATH command. This PATH command instructs DOS to search each directory shown, for example,

> PATH=C:\DOS;C:\BATCH;C:\UTIL

If no path is specified and the program or batch file is not found in the current directory, DOS first looks for the file in the C:\DOS directory. If the file is still not found, DOS then checks the C:\BATCH directory and then the C:\UTIL directory.

If you include more than one directory in the PATH command, you must separate the directories with a semicolon (;). To issue the PATH command at the DOS prompt, type the following:

> **PATH** *d:path specifier;d:path specifier...*

The drive specifier *d:* names the drive on which DOS is to search. The first *path specifier* is the first directory on the search path. The semicolon (;) separates the first directory from the optional second directory. The ellipses shows that you can have other path specifiers in your command line.

DOS takes some time to search the directories in a search path for a program or batch file. DOS searches the directories in the order listed in the PATH command. You should list the directories in the order of most use, therefore. If you usually use DOS commands, place the C:\DOS directory first in the path. As you become more experienced with batch files, you may find that you use batch files more than you use DOS commands. In this case, place the C:\BATCH directory first in the path.

DOS retains the PATH until you change the command or reboot the computer. Because you must specify a search path every time you boot your computer, the PATH command is found in the AUTOEXEC.BAT file of every computer with a hard disk.

Exercise 1.1: Changing the PATH

For this exercise you need a formatted floppy disk.

1. Do a warm boot to restart your computer. Then insert the formatted floppy into drive A. Make sure that the default drive is drive C.

2. Examine the current PATH statement in the AUTOEXEC.BAT file. Type **path** and press ⏎Enter.

3. Type the command

   COPY CAT.TXT A:

   and press ⏎Enter.

   DOS should display the File not found error message on-screen. This message appears because you did not specify the full path in the COPY command and the DOS_WORK directory is not currently in the PATH statement.

4. Temporarily reset the PATH statement. Type

   PATH=A:\;A:\TMP;C:\DOS_WORK;C:

   and press ⏎Enter.

5. Examine the PATH statement, using the PATH command, to make sure that your statement matches the command in step 4. Repeat step 4 if you made a mistake.

6. Now try to copy the CAT.TXT file. Type

   COPY CAT.TXT A:

   and press ⏎Enter.

   Notice this time that the file was found. The file was located because you put the DOS_WORK directory in the path so that DOS would search in that directory.

Issuing a new path statement at the DOS prompt, as you did in step 4, changes the DOS search path for the current computer session only. The path is saved in memory, but not anywhere on disk. Objectives 5 and 6 of this chapter describe how to permanently change the path statement.

Objective 2: To Change the DOS Prompt by Using the PROMPT Command

The DOS prompt is another visible part of your computer system that you can customize. In fact, if you have a hard disk, you almost have to customize the DOS prompt.

The default prompt displays the current drive followed by the greater than sign:

A>

This prompt is useful if you have only floppy disks. If you have a hard disk, however, the default prompt is inadequate. You want the prompt to show the current disk and the current directory. If you have a hard disk, the standard prompt used by most people is the current path and the greater than sign:

C:\DOS>

This prompt has been the standard throughout the book. Now you learn how to use the PROMPT command to customize the prompt. The syntax for the PROMPT command is

PROMPT *text*

where *text* is any combination of words or special characters. The term used to describe the special characters is *metastring*.

Understanding Metastrings

A metastring consists of two characters—the first is the dollar sign ($) and the second is a keyboard character. DOS interprets metastrings to mean something other than the literal character definition. The metastring $t in the PROMPT command, for example, tells DOS, "Display the current time, in the HH:MM:SS.XX format," in the DOS prompt. $t displays the current system time as part of or all the command prompt.

DOS recognizes the symbols greater than (>), less than (<), and vertical bar (|) as special characters. These characters have metastring equivalents, which enable you to use them in PROMPT text. You must substitute an appropriate metastring to cause these special characters to appear in the prompt. Otherwise, DOS tries to act on the characters in its usual way. Table 7.1 summarizes metastring characters and their meanings to the PROMPT command.

Table 7.1 Metastring Characters	
Character	*Produces*
$	$, the actual dollar sign
_ (underscore)	Moves the cursor to the next line
b	Vertical bar (|)
d	Current date
g	Greater than (>) character
l	Less than (<) character
n	Current disk drive letter
p	Current drive and path
q	Equal (=) character
t	System time
v	DOS version
Any other character	Character is ignored

Customizing Your Prompt

Following is the standard prompt command:

    PROMPT pg

The $p displays the current path, including the drive. The $g displays the greater than sign.

You can use the metastring characters with the PROMPT command to produce your own DOS prompt. PROMPT enables you to use words or phrases in addition to metastrings. You can experiment with different combinations of metastrings and phrases.

When you find a combination that you like, type the PROMPT command and the metastring and phrase in your AUTOEXEC.BAT file. Then each time you boot your computer, your custom prompt appears. Issuing the PROMPT command alone with no parameters restores the prompt to its default—the drive name and the greater than sign (C>).

Exercise 2.1: Changing the Prompt

Use table 7.1 to work with commands for changing the prompt statement.

1. Boot your computer, if necessary.
2. Change the prompt so that it reads

   ```
   The Current Path is [path].
   ```
3. Type

   prompt the current path is $p

   and press ⏎Enter. Examine the resulting prompt.
4. Change the prompt string again so that the > appears after the path.
5. Type

   prompt the current path is pg

   and press ⏎Enter.

Again, you have changed the prompt only temporarily, by storing the instruction in RAM. After you reboot the computer, the original prompt reappears.

Objective 3: To Understand Batch Files

A *batch file* is a text file that contains DOS statements. These statements execute DOS commands, execute programs, change the computer environment, or even provide special processing that is possible only in a batch file. DOS executes these statements one line at a time, treating them as though you entered each statement individually.

Batch files always have the extension BAT in their full file names. After you type a batch file's name at the DOS prompt, COMMAND.COM searches the current directory and the search path for a program or a batch file with that file name. COMMAND.COM then reads the batch file and executes the statements that the file contains. The whole process is automatic. You enter the batch file name, and DOS does the work.

You can name a batch file anything you want as long as the name has a BAT extension. After you type a name at the DOS prompt, however, DOS first looks for that name as an internal command built into COMMAND.COM. If you name a batch file COPY.BAT, for example, you can never execute the batch file because DOS will execute the COPY command instead.

If DOS does not find the name in its list of internal commands, DOS looks for a program file with a COM or EXE file extension before the program looks for a file with a BAT extension. For this reason, never use a batch file name that is the same as a DOS command name. If the \DOS directory is before the \BATCH directory in the PATH command, the DOS command executes. If the \BATCH directory is first, the batch file executes. You can avoid this conflict and confusion by using unique names.

Batch files cannot accomplish everything, but they can make using your computer easier and more pleasant. The batch file varies from being the most ignored to being the most abused of all programs. Many books are devoted to the batch file's capabilities—some to a point where readers walk away scratching their heads in confusion. Actually, you can consider the batch file to be the nonprogrammers' programming language. It is a limited yet powerful language: each word can have the power of several lines of DOS commands with the power of programming code.

At the command line, you can review the contents of a batch file by simply typing the TYPE command. In seconds, you can alter a file that changes your PC's personality.

Because you are using the command line, batch files are useful for issuing multiple commands that are complex, potentially destructive, or easy to mistype at the command line. A good example is some form of the BACKUP command with several switches.

BACKUP is a simple command, but it can be confusing because you probably don't use it often. You may find it more convenient to put properly formed backup commands in batch files that you can execute with one simple batch name. This chapter contains a batch file for a full system backup that you can use as presented or change to fit your situation.

Because batch files can display text that you enter, you can compose screens that enable you to execute commands along with syntax examples, reminders, and explanatory notes about the commands. Also, you can simply display a message of the day.

In their advanced forms, batch files resemble programs. This chapter teaches you the basic forms of batch files. If advanced techniques in batch processing interest you, be sure to read about batch files in Que's *MS-DOS 5 User's Guide*, Special Edition.

Objective 4: To Create a Batch File

Batch files contain ASCII text2 characters. Word processors save documents in a special format with special codes for margins; indents; formats such as boldface, type style, and size; and other information. Most programs, however, give you the option of saving the file without these special codes. The word processor may call this unformatted, text, or ASCII format. If you save a formatted file as a batch file, the results may be unpredictable.

DOS now contains a handy, easy-to-use, full-screen text editor called EDIT. If you have heard about the DOS line editor, EDLIN, from prior versions of DOS, forget that EDLIN ever existed. EDLIN was not only difficult to use, but with it you could change only one line at a time.

An easy way to create a very short and simple batch file from the command line is to use the COPY CON command. Just remember that COPY CON does not enable you to correct a text line after you press ⏎Enter.

Storing Batch Files

The best place to store batch files is in a directory named BATCH. Do not put your batch files in the root directory or the DOS directory.

In a way, the root directory is the most important directory on your hard disk because it contains the files needed to boot your computer, and it contains the first level of subdirectories, such as \DOS and your applications' directories. When you view the files in the root directory, you do not want the file list to be complicated by any other files.

Do not put the batch files in the DOS directory because this directory should contain only files that are a part of DOS. When you upgrade to another version of DOS, you do not want to sort through a single directory that contains DOS files and other files to determine the files you want to keep and those you want to upgrade.

Creating a Batch File with COPY CON

You are familiar with the COPY command. CON is a reserved device name in DOS. *CON* stands for *CONSOLE* and refers to the keyboard or the screen. To copy a file from CON means to copy from whatever you type at the keyboard. To copy a file to CON means to display it on-screen.

Using COPY CON, you can create a batch file that clears the screen and presents a wide directory. This batch file automates the CLS command and the DIR command with the /W (wide display) switch. The syntax for COPY CON is

   COPY CON *d:filename.ext*

COPY CON is a handy utility for creating small text files of all kinds and should become a part of your toolbox of DOS tricks.

Exercise 4.1: Creating a Simple Batch File

For this exercise, you need a formatted floppy disk. You will create a batch file that clears the computer screen and displays the date and time, after just two keystrokes from you.

1. Boot your computer, if necessary, and insert a formatted floppy disk into drive A.
2. Make drive A the default drive.
3. At the A:\> prompt, do the following:

   Type **copy con a:dt.bat** and press (↵Enter).

   Type **cls** and press (↵Enter).

   Type **date mm-dd-yy** and press (↵Enter).

   Type **time** and press (↵Enter).
4. Press (Ctrl)+(Z) (hold down the (Ctrl) key and press (Z)) or press the (F6) function key.
5. Press (↵Enter).

   DOS displays the message

   ```
   1 File(s) copied.
   ```
6. To see that the directory contains the new batch file, type

   dir a:dt.bat

   and press (↵Enter).
7. To try out the batch file, type **dt** and press (↵Enter).

The screen clears, and the date and time, according to the computer's system clock, are displayed. You have just created a new DOS batch file, using COPY CON.

At best, use COPY CON to create very short batch files. After you press ⏎Enter to complete a line, you cannot go back and change that line. Most of the time, you should use the DOS Editor to create and modify batch files.

Looking at Rules for Batch Files

When you create batch files, you must follow certain rules. The following list is a summary of those rules:

- Batch files must be ASCII text files. If you use a word processor, make sure that you save the file as a straight text or ASCII file.
- The name of the batch file can contain from one to eight characters. The name must conform to the DOS rules for naming files. The best practice is to use alphabetical characters in batch file names.
- The file name must end with the extension dot (.), followed by the BAT extension.
- The batch file name should not be the same as a program file name (a file with an EXE or COM extension).
- The batch file name should not be the same as an internal DOS command (such as COPY or DATE).
- The batch file can contain any DOS commands that you enter at the DOS prompt.
- You can include program names that you usually type at the DOS prompt.
- Use only one command or program name per line in the batch file.

Looking at Rules for Running Batch Files

You start batch files by typing the batch file name (excluding the extension) at the DOS prompt. The following list summarizes the rules DOS follows when it loads and executes batch files:

- If you do not specify the disk drive name before the batch file name, DOS uses the current drive.
- If you do not specify a directory path, DOS searches through the current directory for the batch file.

7

- If the batch file isn't in the current directory and you didn't precede the batch file name with a directory path, DOS searches the directories in the search path.

- If a batch file has the same name as a program and both are in the same directory, DOS executes the program, not the batch file. If the program and the batch file are in different directories, DOS executes the item it finds first in the search path.

- If DOS finds a syntax error in a batch file command line DOS displays an error message, skips the errant command, and then executes the remaining commands found in the batch file. Depending on the error, DOS may not pause at the error, but may flash an error message too fast to read, and then continue. Do not assume that a batch file is correct because it seemed to end normally. Check to make sure that every command actually worked correctly.

- You can stop a batch command by pressing Ctrl + C or Ctrl + Break. DOS prompts you to confirm that you want to terminate the batch file. If you answer no, DOS skips the current command (the one being carried out) and resumes execution with the next command in the batch file.

If you try to run a batch file and DOS displays an error message, you probably made a mistake when you typed the name. You can view any batch file by using the TYPE *filename* command.

Objective 5: To Understand the AUTOEXEC.BAT File

One batch file has special significance to DOS. The full name of this batch file is AUTOEXEC.BAT. DOS automatically searches for this file in the root directory after you boot your computer. If an AUTOEXEC.BAT file is present, DOS executes the commands in the file.

Technically, the AUTOEXEC.BAT file is optional; however, every computer with a hard disk should have an AUTOEXEC.BAT file. If for no other reason, you need an AUTOEXEC.BAT file so that you can specify a path and prompt every time you boot your computer.

Most users or system managers add an AUTOEXEC.BAT file of their own design on their boot disk. Using a custom AUTOEXEC.BAT file enables you to benefit from commands that automatically launch operating parameters.

You can omit AUTOEXEC.BAT, manually enter the commands you might include in an AUTOEXEC.BAT file, and accomplish the same result as an AUTOEXEC.BAT file. DOS, however, executes the file if it is there, so why not take advantage? As a rule, AUTOEXEC.BAT files are not distributed with the DOS package because different users need different commands.

During the automated installation, DOS creates, verifies, or changes your AUTOEXEC.BAT file to include a path to the \DOS directory plus any other commands it needs to run on your computer. Some application programs come with installation programs that create or modify AUTOEXEC.BAT as an installation step of the package's main program.

If you have doubts about which commands to include in your AUTOEXEC.BAT file, the following sections may give you some ideas. You can include any commands you want in the AUTOEXEC.BAT file. Decide what you want the AUTOEXEC.BAT file to do, and follow certain rules. The following list is a summary of these rules:

- The full file name must be AUTOEXEC.BAT, and the file must reside in the root directory of the boot disk.
- The contents of the AUTOEXEC.BAT file must conform to the rules for creating any batch file.
- When DOS executes AUTOEXEC.BAT after a boot, you are not prompted for the date and time automatically. You must include the DATE and TIME commands in your AUTOEXEC.BAT file if your computer does not have a built-in clock with a battery backup (only the oldest PCs do not have a clock).

Using AUTOEXEC.BAT is an excellent way for you to set up system defaults. That is, AUTOEXEC.BAT is the place to put commands you would want to enter every time you start your system. You can use AUTOEXEC.BAT, for example, to tell your computer to change to the directory that holds your most commonly used program and then to start that program. Used this way, AUTOEXEC.BAT runs your program as soon as you boot your computer.

Figure 7.1 shows an example of an AUTOEXEC.BAT file—a batch file whose commands DOS executes each time you boot the computer.

Table 7.2 lists the commands most frequently included in simple AUTOEXEC.BAT files plus a few other commands to give you an idea of what you may add to your file. The table also explains each line of the batch file shown in figure 7.1.

7

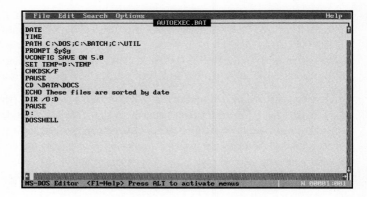

Fig. 7.1
An
AUTOEXEC.BAT
file.

Table 7.2 AUTOEXEC.BAT File Commands

Command	*Function in the AUTOEXEC.BAT File*
DATE	Sets the computer's clock to set up the correct date so that DOS can accurately "date stamp" new and modified files. Most computers have a built-in clock and do not need this command.
TIME	Sets the computer's clock to establish the correct time so that DOS can accurately "time stamp" new and modified files. The DATE and TIME commands also provide the actual date and time to programs that use the computer's internal clock. Most computers have a built-in clock and do not need this command.
PATH	Tells DOS to search the named subdirectories for files that have EXE, COM, or BAT extensions.
PROMPT	Customizes the system prompt. The DOS prompt configuration can include information that makes navigating in directories easier. If you use the PROMPT command in the AUTOEXEC.BAT file, you don't need to enter the optional parameters each time you boot.

Command	Function in the AUTOEXEC.BAT File
VCONFIG	Turns on a screen saver that automatically blanks the screen after five minutes of inactivity. This is *not* a typical command found in an AUTOEXEC.BAT file. This is a program that comes with a certain brand of video adapter card. This command is included here to demonstrate that each AUTOEXEC.BAT file can contain special commands that fit each computer's environment. The more complicated your computer environment, the more of these special commands you are likely to have.
SET TEMP	Stores temporary files. DOS, Windows, and other programs need a place on the hard disk to store temporary files. If you do not have a SET TEMP command in your AUTOEXEC.BAT file, DOS adds it during installation.
CHKDSK/F	Verifies that there are no errors in the directory or file allocation table on your boot disk. Many people like to verify this each time they boot.
PAUSE	Pauses until you press a key so that you have a chance to see the result of the CHKDSK before the next command executes. Pause is used again after the DIR command for the same reason.
CD \DATA\DOCS	Changes the current path. If you have a data directory that you use most often, you may want to check it each time you boot.
ECHO	Enables you to include a message as part of your start-up when used in the AUTOEXEC.BAT file. On a floppy disk system, this message can remind you to insert a program disk into drive A.
DIR /O:D	Lists all the files in the \DATA\DOCS directory in date order
D:	Changes to another disk drive
DOSSHELL	Starts the Shell automatically when you boot

7

Objective 6: To Create an AUTOEXEC.BAT File

The AUTOEXEC.BAT file is a privileged batch file because DOS executes its batch of commands each time you boot your computer. In every other sense, however, AUTOEXEC.BAT is like any other batch file. The best way to create and modify an AUTOEXEC.BAT file is with a text editor or word processor.

Exercise 6.1: Viewing the AUTOEXEC.BAT File

You can see whether AUTOEXEC.BAT exists in your root directory or on your logged floppy disk.

1. Boot your computer, if necessary.
2. View your AUTOEXEC.BAT file by using the TYPE command. Type

    type autoexec.bat

    and press ⏎Enter.

How does your computer's AUTOEXEC.BAT file differ from the one shown in figure 7.1?

Printing or writing down the contents of the AUTOEXEC.BAT file before you make any changes is a good idea. Copy the syntax correctly. This copy will serve as your worksheet. You can use your copy of AUTOEXEC.BAT to find out whether a PROMPT or PATH command is contained in the batch file. If you want to add or alter PROMPT or PATH commands, jot the additions or changes on your worksheet. Use your paper copy of the AUTOEXEC.BAT file to check for proper syntax in the lines you change or add before you commit the changes to disk.

Backing Up the Existing File

Always make a backup copy of your existing AUTOEXEC.BAT file before you make any changes to the file or install any programs. Remember that some program installation procedures automatically change your AUTOEXEC.BAT file. Save the current version by renaming it with a different extension.

Keeping Several Versions of AUTOEXEC

Technically speaking, your computer can have only one AUTOEXEC.BAT file. You can benefit from having several "versions" on hand by giving different extensions to files named AUTOEXEC. You then can activate an alternative version by using the COPY command.

You can use in the extensions any character that DOS normally enables you to use in file names. The extensions NEW, TMP, and 001 are just a few examples. By giving an AUTOEXEC file a unique name, such as AUTOEXEC.TMP, you can activate any other AUTOEXEC file by copying it to AUTOEXEC.BAT. To make AUTOEXEC.TMP your current AUTOEXEC file, for example, type

    copy autoexec.tmp autoexec.bat

Then, to restore your "normal" AUTOEXEC.BAT file, copy your archive copy to AUTOEXEC.BAT by typing the following:

    copy autoexec.arc autoexec.bat

This method is handy if you want to include commands for special activities, such as automatically starting a monthly spreadsheet. When the monthly work is done and you no longer need the spreadsheet when you boot, you can reactivate your normal AUTOEXEC file.

7

Objective 7: To Understand the CONFIG.SYS File

AUTOEXEC.BAT is not the only file that DOS looks for when you boot your computer. Before DOS reads your AUTOEXEC.BAT file—in fact, before DOS even loads the command processor—it looks for the CONFIG.SYS file.

CONFIG.SYS is DOS's system configuration file. Not only does DOS provide built-in services for disks and other hardware, but DOS also extends its services for add-on hardware. The additional instructions that DOS needs to incorporate such outside services as a mouse and other devices are included in the CONFIG.SYS file.

CONFIG.SYS is also the location for naming the values of DOS configuration items that can be "tuned." Files and buffers, discussed in the next section, are two "tunable" DOS items. CONFIG.SYS is a text file like AUTOEXEC.BAT that you can display on-screen or print. You also can change the contents of CONFIG.SYS by using the DOS Editor.

DOS does not execute CONFIG.SYS as it does AUTOEXEC.BAT. Instead, DOS reads the values in the file and configures your computer to agree with those values. Many software packages modify or add a CONFIG.SYS file to the root directory. The range of possible values in the file is wide, but some common values that you can include do exist.

Specifying Files and Buffers

When DOS moves data to and from disks, it does so in the most efficient manner possible. For each file that DOS acts on, an area of system RAM helps DOS track the operating details of that file. The FILES command in CONFIG.SYS controls the number of built-in RAM areas for this tracking operation. (A command establishes in the CONFIG.SYS file the value of a system setting that can be modified.) If a program tries to open more files than the FILES command setting supports, DOS tells you that too many files are open.

Do not be tempted to set your FILES command to a large number just so that you always have room for more open files. The system memory for programs is reduced by each extra file included in FILES. As a rule of thumb, a safe compromise is 30 open files. To set the number of open files to 30, type **files=30**.

The installation documentation for many programs tells you the minimum number of files the program needs. Make sure that the FILES command is at least as large as the largest number a program requires. If one program requires 20 and another suggests 30, for example, use **FILES=30**. Do not add the numbers together and type FILES=50.

Similar to the FILES command is the BUFFERS command. *Buffers* are holding areas in RAM that store information coming from or going to disk files. To make the disk operation more efficient, DOS stores disk information in file buffers in RAM. DOS then uses RAM, rather than the disk drives, for input and output whenever possible.

If the file information needed is not already in the buffer, new information is read into the buffer from the disk file. The information that DOS reads includes the needed information and as much additional file information as the buffer will hold. With this buffer of needed information, DOS probably can avoid constant disk access. The principle is similar to the way a mechanic might use a small tool pouch. Holding frequently used tools in a small pouch relieves him of having to make repeated trips across the garage to get tools from his main tool chest.

Like the FILES command, however, setting the BUFFERS command too high takes needed RAM away from programs and dedicates it to the buffers. The optimum number of buffers depends on the size of your hard disk and type of application:

Hard Disk Size	Number of Buffers
Less than 40M	20
40M–79M	30
80M–119M	40
120M or more	50

Creating and Changing a CONFIG.SYS File

Because the CONFIG.SYS file is a text file, you can use the same methods to create, change, archive, and copy this file as you use with the AUTOEXEC.BAT file. Use the Editor to create and change the CONFIG.SYS file. Make sure that you have a backup copy and an extra copy on a floppy disk before making changes. Note that errors in your CONFIG.SYS file cause your computer to hang more often than an error in the AUTOEXEC.BAT file, making having a boot floppy disk handy before you start making changes to the CONFIG.SYS file even more important.

Figure 7.2 shows a sample CONFIG.SYS file for a 386 computer with a 40M hard disk.

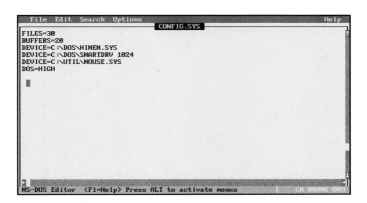

Fig. 7.2
A CONFIG.SYS
file.

Your CONFIG.SYS file may contain other device drivers and DOS settings. Because the configuration of each computer may vary, the exact contents of the CONFIG.SYS file will vary. If you want to explore configurations and device concepts in more detail, consult Que's *MS-DOS 5 User's Guide*, Special Edition. If you are unsure about the device drivers for new peripherals you buy, ask your dealer to explain how to incorporate the new device.

Chapter Summary

In this chapter, you touched the tip of the iceberg of customization options for your computing environment. Entire books are devoted to this topic. Your computer can be a much friendlier machine if you configure it so that it works with you and for you, rather than against you. The AUTOEXEC.BAT file and other batch files that you create can be powerful tools for efficient computing. Always keep in mind the necessity of maintaining proper backups of these important files!

Testing Your Knowledge

True/False Questions

1. You can view the path statement by typing **path** at the DOS prompt or by viewing the contents of the AUTOEXEC.BAT file.
2. $P, $G, and $T are examples of metastrings.
3. Batch files must end with the file extension BAT in order to run.
4. The AUTOEXEC.BAT file comes with most computer systems.
5. You should set the file and buffer size in the CONFIG.SYS file as high as it will go.

Multiple Choice Questions

1. To tell DOS to search, in order, the root directory of drive C, the root directory of drive A, and the DOS subdirectory of drive C, use the command

A. PROMPT C:;A:;C:\DOS

B. PATH PG

C. PATH A:\; C:\;DOS;

D. PATH C:\;A:\;C:\DOS

E. none of the above

2. The result of the command PROMPT TimeQ$T_$P$G is most accurately described by

A. Time=HH:MM:SS.XX

   C:\DOS>

B. TIME=HH:MM:SS.XX_C:\DOS

C. HH:MM:SS.XX C:

D. TIME_C:\DOS>

E. none of the above

3. To convert characters you type on the keyboard directly into a file, you can use the command

A. FILE

B. TYPE

C. COPY CON

D. CON COPY

E. CONSOLE

4. Each time you boot the computer, the

A. AUTOEXEC.BAT file is executed.

B. CONFIG.SYS file is executed.

C. DOS file is executed.

D. KEYBOARD.CFG file is executed.

E. none of the above

5. When creating batch files, remember that

A. batch files must be ASCII text files.

B. batch file names must end with the extension BAT.

C. batch files cannot have multiple commands per line.

D. batch files should not be the same as program file names.

E. all of the above (all are valid rules)

Fill-In-the-Blank Questions

1. In the path statement, path specifiers must be separated by a(n)
   _____ *se mi-colon* _____ .

2. Path and prompt statements are generally stored in the file
   _____ *AUTOEXEC. BAT* _____

3. The statement PROMPT PG is necessary for use with a(n)
   _____ *hard* _____ disk.

4. Batch files should never reside in the _____ *Root* _____ directory.

5. CONFIG.SYS should always reside in the _____ *Root* _____
   directory.

Review: Short Projects

1. Printing the AUTOEXEC.BAT File

   Print the AUTOEXEC.BAT file of your computer. For each line of text,
   write an explanation of what each command does. Give the printout to
   your instructor.

2. Printing the CONFIG.SYS File

   Print the CONFIG.SYS file of your computer. For each line of text,
   write an explanation of what each command does. Give this printout
   to your instructor.

3. Planning a Batch File

   On a piece of paper, write the contents of a batch file that will do the
   following things:

   A. Set the prompt so that the drive, path, and > will appear. *Prompt PG*

   B. Set the path so that the root, DOS, and batch directories are
      searched. *Path c:\DOS; c:\Dos\Bat*
      c:\DOS\UTIL

   C. Clear the screen. *CLS.*

   D. Print the current date and time on-screen. *DATE* *Time >PRN*

.BAT.
.TXT.
.EXE.
.COM.

Review: Long Projects

1. Creating a Batch File

   On a formatted floppy disk, create a batch file called AUTO.BAT that
   contains the instruction lines you wrote for Short Project #3 of this
   chapter. Run the file to make sure that it works correctly.

   copy con. con.

7

2. Creating Your Own Batch File

   Write a batch file that starts a computer program you use all the time. Use the following example, which starts the game TETRIS from the GAMES subdirectory of drive C, and returns the user to the root directory of drive C after the user exits the game.

```
C:
C\GAMES
TETRIS
CD\
```

BAT   CD

Setup and Installation

This appendix tells you how to install or upgrade to MS-DOS 5. Why put this at the end of the book if it's the first thing you must do? The answer is simple. Many computers are sold today with DOS already installed on the hard disk. Also, if you are using a computer at work, you may have an Information Center or PC Support Group that installs DOS and other software on your hard disk for you.

If DOS is installed on your computer, you can skip this appendix for now. Go right to Chapter 1, and learn about MS-DOS 5 from the beginning. You may need this information later to help you upgrade to a new version of DOS.

Key Terms in This Appendix	
Low-level format	The process that physically prepares a hard disk so that it can be used by DOS. This process is not the same as the DOS FORMAT command and is not a part of DOS.
Boot	To start your computer and load DOS into the computer's memory.
Partition	A section of a hard disk set up so that it can be used by DOS. A partition can be all or part of a hard disk. Each partition has a separate drive letter and is considered a separate disk by DOS.
Working disks	Copies of the original DOS disks. If anything should happen to your working disks, you can re-create them from your original DOS disks.
Uninstall	The process of reverting to your prior version of DOS if you have a problem after you upgrade to MS-DOS 5.

A

Requirements To Install MS-DOS 5

MS-DOS 5 can run on any PC with at least 256K of memory. Virtually every PC sold since 1983 has at least this much memory. Computers sold since 1986 usually have the maximum of 640K of conventional memory. If you plan to install MS-DOS 5 on a hard disk, you need at least 512K of memory and 2.5M of available disk space.

If you are upgrading from a previous version of DOS, execute CHKDSK to determine the amount of available disk space (see fig. A.1). If less than 2.5M are available, use COPY or BACKUP to copy some of or all the files on the hard disk to floppy disks. Then delete enough files from the hard disk so that at least 2.5M are available.

If you plan to install MS-DOS 5 on floppy disks, you need a supply of blank disks that fit into drive A. If drive A is a 5 1/4-inch drive, for example, you need seven 360K floppy disks. If drive A is a 3 1/2-inch drive, you need four 720K disks.

```
C:\>chkdsk

Volume DRIVE_C      created 10-03-1990 3:45p
Volume Serial Number is 166D-BDE3

  27553792 bytes total disk space
     75776 bytes in 3 hidden files
    112640 bytes in 48 directories
  24080384 bytes in 920 user files
     20480 bytes in bad sectors
   3264512 bytes available on disk

      2048 bytes in each allocation unit
     13454 total allocation units on disk
      1594 available allocation units on disk

    655360 total bytes memory
    605232 bytes free

C:\>
```

Fig. A.1
The CHKDSK
screen.

Making Backup Copies

If your computer has a previous version of DOS installed, or you have another computer available with any version of DOS installed, your first step is to make backup copies of the DOS disks.

If you are installing DOS on your computer for the first time and you have no other computers available, you cannot make backup copies until you complete the installation. In this case, you may want to ask your computer dealer to make backup copies of the DOS disks for you or to install DOS on your hard disk for you.

If you cannot copy the DOS disks and cannot successfully complete the installation because you get errors reading the DOS disks, contact your computer dealer immediately and get replacement disks.

To make backup copies of the DOS disks, you need five or six 5 1/4-inch 360K floppy disks or three 3 1/2-inch 720K disks. If you have two floppy disk drives that are the same size, follow these instructions:

1. Type **diskcopy a: b:/v**, and press ↵Enter.
2. At the prompt to insert the SOURCE disk into drive A, put the original MS-DOS 5 disk labeled "Disk 1" into drive A.
3. You are prompted to insert the TARGET disk into drive B.

   If you have two 5 1/4-inch drives, put a blank 360K floppy disk into drive B, close the drive door, and press ↵Enter. This disk does not have to be formatted.

MS-DOS SmartStart

If you have two 3 1/2-inch drives, put a blank 720K disk into drive B and press ⏎Enter. This disk does not have to be formatted.

4. When the DISKCOPY process is complete, remove the disks from both drives and label the new disk "MS-DOS 5—Disk 1."

5. At the prompt Copy another disk (Y/N)? press Y, then ⏎Enter, and repeat steps 2 through 4 for each original MS-DOS 5 disk.

6. When you complete the DISKCOPY for the last original MS-DOS 5 disk, you are asked whether you want to copy another disk. Press N and then ⏎Enter.

If you have only one floppy disk drive or drive A and drive B are different sizes, follow these steps:

1. Type **diskcopy a: a:/v**, and press ⏎Enter.

2. When you are prompted, place the original MS-DOS 5 disk labeled "Disk 1" into drive A, and press ⏎Enter.

3. At the prompt to insert the TARGET disk into drive A, remove the original MS-DOS 5 disk, and insert a blank 360K or 720K disk into drive A; then press ⏎Enter. This disk does not have to be formatted.

4. You may be prompted to remove the TARGET disk and insert the SOURCE disk again. You are then prompted to insert the TARGET disk again. Switch the same two disks as prompted until the DISKCOPY completes.

5. When the DISKCOPY process is complete, remove the disk from drive A, and label the new disk "MS-DOS 5—Disk 1."

6. When you complete the DISKCOPY for the last original MS-DOS 5 disk, you are asked whether you want to copy another disk. Press Y; then press ⏎Enter. Repeat steps 2 through 5 for each original MS-DOS 5 disk.

7. When you complete the DISKCOPY for the last original MS-DOS 5 disk, press N; then press ⏎Enter.

If the DISKCOPY fails, repeat the process using a different blank disk. If you cannot successfully copy and original MS-DOS 5 disk, contact your dealer for a replacement disk.

Store the original MS-DOS 5 disks in a safe place, and use the copies. If you get any errors using a copy of one of the DOS disks, repeat the DISKCOPY process to re-create that disk. If you cannot make a copy of the disk that you can use for the install, contact your dealer for a replacement.

In all subsequent instructions, any reference to an original DOS disk refers to the copy of the original that you just made. If you are installing DOS for the first time and do not have copies, you must use the original disks.

Installing MS-DOS 5

If you are installing MS-DOS 5 on a hard disk that already contains a copy of DOS Version 2.11 or later, skip this section, and read "Upgrading from a Previous Version of DOS." If you have a new hard disk or a hard disk with a DOS version prior to 2.11, you should install MS-DOS 5 as if your hard disk did not contain DOS.

If you are not upgrading a previous version of DOS, make sure that you have the MS-DOS 5 OEM package. If you have the MS-DOS 5 Upgrade package, you will not be able to boot your computer.

Installing on a Hard Disk

This section describes how to install MS-DOS 5 on a hard disk that does not contain a version of DOS. You have many options and possible configurations with a new hard disk; this book covers the basic configuration used by most people. If you plan to install multiple partitions or you have multiple hard disks, consult *Que's MS-DOS 5 User's Guide*, Special Edition, or the DOS manual.

Before DOS can use a hard disk, the disk must be low-level formatted. Low-level formatting is not the same as the DOS FORMAT command. A low-level formatting program is not supplied with DOS. When you buy a new hard disk or a new computer with an installed hard disk, the dealer usually has formatted the disk. In some cases, the dealer supplies a separate program to low-level format the drive. If you cannot install DOS because DOS does not recognize the hard disk, the disk has not been low-level formatted. If you do not have a low-level formatting program, contact your dealer.

To install DOS on a hard disk, follow these instructions:

1. Place Disk 1 into drive A. (Close the drive door if applicable.)
2. Turn on your computer. If your computer is on, hold down the Ctrl and Alt keys and press the Del key; then release all three keys. This procedure boots, or restarts, the computer. DOS loads and starts the SETUP procedure.

3. The SETUP procedure checks your hardware and starts the installation process. From time to time, DOS asks you for the following
   information:

   > The date and time
   >
   > The country
   >
   > The keyboard type
   >
   > Whether to install DOS on a hard disk or floppy disks
   >
   > For hard disk installation, which directory to use for the DOS files

   For most of this information, SETUP offers a default response. Usually,
   you can accept the defaults for U.S. installation with a U.S. keyboard.
   For hard disk installation, DOS is installed on drive C in the C:\DOS
   directory by default.

   DOS also gives you the option to continue or cancel the installation
   any time during the installation process. If you cannot complete the
   installation for any reason, you can start from the beginning at a later
   time.

4. When you are prompted to insert a different disk, remove the disk
   from drive A, insert the requested disk, close the drive door if appropriate, and press ⏎Enter. If you insert the wrong disk, DOS tells you
   and gives you a chance to insert the correct disk.

5. Before DOS can use a hard disk, it must establish the disk as a DOS
   disk and assign it a logical drive letter, such as drive C. If the drive
   has not been partitioned, SETUP runs the FDISK program to partition
   the disk. Accept the defaults to use the entire disk as a single DOS
   partition.

6. After the disk is partitioned, SETUP formats the disk. This format
   establishes the disk directory and File Allocation Table. The DOS
   FORMAT command is explained in Chapter 5. After the format is
   complete, you can give the disk a volume label of 1 to 11 characters,
   or press ⏎Enter to not give the hard disk a label.

7. SETUP then copies the DOS files on to the hard disk. SETUP displays
   a horizontal bar that shows what percentage of the total installation
   process is complete. SETUP displays the current activity in the lower
   right corner of the screen.

8. When the installation process is complete, remove the disk from
   drive A, and hold down the Ctrl and Alt keys and press the Del;
   then release all three keys. This process reboots or restarts the computer using the files on the hard disk.

The installation process on the hard disk is complete.

Installing on Floppy Disks

If you do not have a hard disk, you can install MS-DOS 5 on floppy disks. If drive A is a 5 1/4-inch drive, you need seven 360K disks. If drive A is a 3 1/4-inch drive, you need four 720K disks. These disks can be unformatted.

If you have a 5 1/4 inch drive, label the 360K disks as follows:

> Startup
>
> Support
>
> Shell
>
> Help
>
> Edit
>
> Utility
>
> Supplemental

If you have a 3 1/2-inch drive, label the 720K disks as follows:

> Startup/Support
>
> Shell/Help
>
> Basic/Edit/Utility
>
> Supplemental

These disks are your operating disks. You will use these disks to run DOS. After you complete this process, store the original DOS disks in a safe place and use the operating disks.

To install DOS on floppy disks, follow these steps:

1. Place Disk 1 into drive A. (Close the drive door, if applicable.)
2. Turn on your computer. If your computer is on, hold down the Ctrl and Alt keys and press the Del key; then release all three keys. This process boots, or restarts, the computer. DOS loads and starts the SETUP procedure.

3. The SETUP procedure checks your hardware and starts the installation process. From time to time, DOS asks you for the following information:

   The date and time

   The country

   The keyboard type

   Whether to install DOS on a hard disk or floppy disk

   For most of this information, SETUP offers a default response. If you are using a U.S. keyboard, you usually can accept the defaults for U.S. installation. Respond to install DOS on floppy drives.

   DOS also gives you the option to continue or cancel the installation any time during the process. If you cannot complete the installation for any reason, you can start from the beginning at a later time.

4. When you are prompted to insert a different disk, remove the disk from drive A and insert the requested disk, close the drive door if appropriate, and press ⏎Enter. If you insert the wrong original DOS disk, DOS gives you a chance to insert the correct disk.

5. SETUP prompts you to insert each of the operating DOS disks that you labeled previously. The program then copies the DOS files on to each floppy disk. SETUP displays a horizontal bar that shows what percentage of the total installation process is complete. SETUP displays the current activity in the lower right corner of the screen.

6. When the installation process is complete, remove the disk from drive A, and insert the operating Setup disk. Hold down the Ctrl and Alt keys and press the Del key; then release all three keys. This process reboots, or restarts, the computer using the new working DOS disk.

The installation process on floppy disks is complete.

Upgrading from a Previous Version of DOS

If you already have DOS Version 2.11 or greater installed on your hard disk, you can upgrade to MS-DOS 5 without having to partition or format your hard disk.

If you have a version of DOS prior to Version 2.11, you cannot upgrade. You must back up your entire hard disk, except for DOS files, and then install MS-DOS 5 using the procedures in the preceding section. Then restore your files to the new hard disk. BACKUP and RESTORE procedures are discussed in Chapter 5.

This section describes how to upgrade a DOS disk with a previous version of DOS to MS-DOS 5. If you have a nonstandard disk that requires special device drivers in CONFIG.SYS, these procedures may not work. Also, you may want to change partition sizes if you have been using a version of DOS prior to DOS 4.0 and you have a hard disk bigger than 32M. In these cases, consult *Que's MS-DOS 5 User's Guide,* Special Edition, or the DOS manual for more detailed technical information beyond the scope of this book.

If you want to upgrade, make sure that you purchase the MS-DOS 5 Upgrade package. This software package differs from the MS-DOS 5 package for initial installation. If you have an existing hard disk with files that you want to keep, make sure that you have the correct package. If the DOS disks are not labeled "Upgrade," contact your dealer before you attempt to upgrade your hard disk.

The procedure to upgrade your hard disk from a previous version of DOS is similar to the procedure to install MS-DOS 5 on a new disk. Some significant differences do exist, however. When you upgrade, you have data on your hard disk that you do not want to lose. Part of the upgrade procedure is to back up all the files on your hard disk and save your old version of DOS.

Before you start the upgrade, make sure that you have one or two blank floppy disks that fit into drive A. This disk or disks serve as uninstall disks and enable you to revert to your existing version of DOS if you have problems after you install MS-DOS 5. If you have a 5 1/4-inch drive A, you need either two 360K disks or one 1.2M disk. If you have a 3 1/2-inch drive A, you need either one 720K or one 1.44M disk. If you have one disk, label it "Uninstall." If you have two 360K disks, label them "Uninstall #1" and "Uninstall #2." Uninstall disks may be blank, formatted disks or unformatted disks.

To upgrade your hard disk, follow these steps:

1. Copy your AUTOEXEC.BAT and CONFIG.SYS files to a floppy disk. You may need them later.

2. Quit any Shell programs or task-switching programs, such as DOS Shell, Windows, DESQView, or Software Carousel.

3. Place Disk 1 into drive A, and close the drive door, if applicable.

4. Type **a:setup**, and press ⏎Enter.

5. The SETUP procedure checks your hardware and starts the installation. You are asked whether you want to back up the data on your hard disk before you upgrade. If you have a complete backup, you can press N. If you do not have a complete backup, press Y.

6. DOS estimates how many floppy disks you need for the backup. Make sure that you have sufficient floppy disks, and follow the prompts to insert each backup disk until the entire disk is backed up.

7. When the backup is complete, you can continue the upgrade or exit. Unless you encounter a problem with the backup procedure, continue with the setup.

8. As the upgrade proceeds, you will be prompted for certain information, such as in which directory to put the MS-DOS 5 files. By default, DOS is placed on drive C in the C:\DOS directory.

9. When prompted, insert the "Uninstall" disk or disks into drive A and press ⏎Enter.

10. When you are prompted to insert a different disk, remove the disk from drive A, insert the requested disk, close the drive door if appropriate, and press ⏎Enter. If you insert the wrong DOS disk, DOS tells you and gives you a chance to insert the correct disk.

    During the upgrade, SETUP renames your AUTOEXEC.BAT file AUTOEXEC.OLD and renames your CONFIG.SYS file CONFIG.OLD. SETUP also renames the directory that contains your existing DOS files to OLD_DOS.1. (If you already have an OLD_DOS.1, SETUP renames the directory OLD_DOS.2, and so on.)

    SETUP then copies the DOS files to the hard disk. SETUP displays a horizontal bar that shows what percentage of the total installation process is complete. SETUP displays the current activity in the lower right corner of the screen.

11. SETUP asks you whether you want to start in the DOS Shell when you boot your computer. It is a good idea to learn to use the DOS Shell first. If you reply Yes, SETUP adds the statement to your AUTOEXEC.BAT file. SETUP may make other changes to your AUTOEXEC.BAT or CONFIG.SYS files.

12. When the install process is complete, remove the disk from drive A, and hold down the Ctrl and Alt keys and press the Del key; then release all three keys. This process reboots, or restarts, the computer using the files on the hard disk.

The upgrade process on the hard disk is complete.

Upgrading on Floppy Disks

You really do not "upgrade" your existing floppy disks to MS-DOS 5. You create a new set of DOS floppy disks. You can install MS-DOS 5 on floppy disks with either the MS-DOS 5 OEM package or the MS-DOS 5 Upgrade package.

If you have the Upgrade package, first boot your computer with your current version of DOS. Then insert Disk 1 into drive A, and type **a:setup /f** and press ⏎Enter. Then follow the instructions for "Installing on Floppy Disks." If you have the OEM package, follow the instructions for "Installing on Floppy Disks."

After the Installation or Upgrade Is Complete

In most cases, after you complete the installation or the upgrade, you are ready to use your computer. In some cases, however, you must do some additional work so that you can improve the way DOS runs.

In the directory in which you installed DOS, such as C:\DOS, the README.TXT file contains additional information about DOS and the installation process.

If you have a 286, 386, or 486 computer with extended memory, read the README.TXT file (see fig. A.2). You may have to modify the HIMEM command if you have one of the following computers:

A

IBM PS/2

A computer with a Phoenix Cascade BIOS

HP "Classic" Vectra

AT&T 6300 Plus

Acer 1100

Toshiba 1600 or 1200 XE

WYSE 12.5 MHz 286 with micro channel

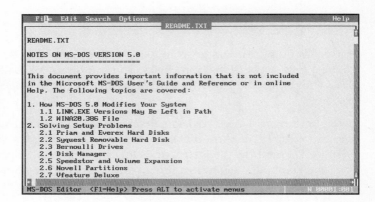

Fig. A.2
The
README.TXT file
contains techni-
cal information
about installing
or upgrading
MS-DOS 5.

To see the README file from the DOS Shell with a mouse, follow these steps:

1. Point to the C:\DOS directory in the Directory Tree area in the upper left portion of the display, and click the left mouse button.

2. Point to the Editor in the Program List area in the lower left portion of the display, and double-click the left mouse button. If nothing happens, press ⏎Enter.

3. In the dialog box, type **readme.txt**, and press ⏎Enter.

4. Read the information in the README.TXT file. Press PgUp and PgDn to see the entire file.

5. To leave the Editor, move the mouse pointer to the File menu at the top of the display, and click the left mouse button. Then move the mouse pointer to the Exit menu choice, and click the left mouse button.

To see the README file from the DOS Shell with the keyboard, follow these steps:

1. Press Tab⇄ to highlight the Directory Tree area in the upper left portion of the display. Press ↓ and ↑ to highlight the \DOS directory.

2. Press Tab⇄ to highlight the Program List area in the lower left portion of the display. Press the ↓ and ↑ to highlight the Editor in the Program List area and press ⏎Enter.

3. In the dialog box, type **readme.txt** and press ⏎Enter.

4. Read the information in the README.TXT file. Press PgUp and PgDn to see the entire file.

5. To leave the Editor, press Alt, F for File, and X for Exit.

To see the README file from the DOS prompt, follow these steps:

1. Type **cd\dos** and press ⏎Enter.
2. Type **edit readme.txt** and press ⏎Enter.
3. Read the information in the README.TXT file. Press PgUp and PgDn to see the entire file.
4. To leave the Editor, press Alt, **F** for File, and **X** for Exit.

This file contains additional technical information about the installation that is outside the scope of this book.

If You Have Problems with MS-DOS 5

If you have upgraded to MS-DOS 5 and cannot operate your computer for any reason, you can revert to the previous version of DOS. This process is called uninstalling DOS 5. Do not try to uninstall if you used MS-DOS 5 to repartition or reformat your hard disk.

To uninstall DOS 5, insert the Uninstall (or Uninstall #1) disk that DOS created during the upgrade and hold down the Ctrl and Alt keys and press the Del key; then release all three keys to reboot your computer.

Follow the prompts to tell DOS to uninstall MS-DOS 5 and revert to your previous version of DOS.

Preserving DOS Shell Settings

You may need to reinstall MS-DOS 5 even if the installation was completed without error. With any of the following situations, you may find it easier to reinstall DOS. You may have a marginal sector on your disk that goes bad, for example, causing part of DOS to be unreadable. If you change your video display, from EGA to VGA, you must follow a complex procedure to upgrade the DOS installation. You will find it much easier to reinstall MS-DOS 5.

Another reason to install DOS again is to upgrade to future new version of MS-DOS 5. If you changed the DOS Shell settings or added program groups, this information is stored in a file named DOSSHELL.INI. After you reinstall MS-DOS 5 or install a future version of DOS 5, the DOSSHELL.INI file in the \DOS directory contains the original default settings. If the DOS Shell starts automatically when you boot DOS, you can see that you lose any customizing or program groups. To recover the custom settings, you must copy your customized DOSSHELL.INI file to the \DOS directory.

A

If you are in the DOS Shell, press F3 to cancel the Shell and return to the DOS command line. Then copy DOSSHELL.INI from the old DOS directory to the \DOS directory.

If your old DOS directory was renamed OLD_DOS.2 during the upgrade and your new DOS directory is DOS, enter the following command:

copy\old_dos.2\dosshell.ini\dos

and press ⏎Enter.

If your old DOS directory had a different name, substitute that name for OLD_DOS.2 in the command.

To return to the DOS Shell, type **DOSSHELL** and press ⏎Enter. The DOS Shell now reflects the changes that you made to the Shell in the previous version of MS-DOS 5.

A

Errors Great and Small

One of the standard routines of slapstick humor involves slipping on a banana peel. What makes this funny? Certainly not the fall. Everyone falls at one time or another, often with unpleasant results. Maybe the humor in slapstick is possible because no one really gets hurt; you know that the actors don't injure themselves, so you can sit back and enjoy the physical levity.

Before you finish this appendix, you will realize that although some of the DOS error messages are no laughing matter, you probably will never see *those* error messages on your screen. In fact, most of the common error messages that you encounter will cause you to smile at the mistakes you can correct easily.

What Are Error Messages?

Ending a book with an appendix on errors may make you wince, but don't lose heart. You can treat most of the common DOS error messages with a grin. Error messages are DOS's way of telling you that you (or DOS) made a correctable blunder. Error messages serve as reminders that both man and machine make mistakes.

Occasionally, you may issue a harmless but incorrect command or forget to close the door on a disk drive. DOS quickly points out such mistakes, but you can correct and learn from them.

How Serious Are Error Messages?

Some error messages indicate serious problems. Fortunately, you probably will never see one of those messages. Also, some potentially serious error messages can be caused by any of several problems; if you remain calm, you might discover an easy remedy.

Read through the following error messages. These messages are the ones that routinely appear on-screen. Disasters are rare, so enjoy your PC, confident that you have earned your wings.

Interpreting Error Messages

If you see a message that you cannot locate in this guide, refer to your computer's MS-DOS manual.

Some messages appear when you start MS-DOS, and some appear while you use your computer. Most start-up errors mean that MS-DOS did not start and that you must reboot the system. Most of the other error messages mean that MS-DOS terminated (aborted) the program and returned to the system prompt. For easy reference, the messages are listed in alphabetical order with explanations of their most common causes.

B

Access denied

You or a program attempted to change or erase a file that is marked as read-only or that is in use. If the file is marked as read-only, you can change the read-only attribute with the ATTRIB command.

Allocation error, size adjusted

This error is a warning message. The contents of a file have been truncated because the size indicated in the directory is not large enough for the amount of data in this file. To correct this problem, use the CHKDSK /F command.

`APPEND/ASSIGN Conflict`

A DOS warning message. You cannot use APPEND on an assigned drive. Cancel the drive assignment before using APPEND with this drive.

`Bad command or filename`

The name you entered is not valid for invoking a command, program, or batch file. The most frequent causes are the following:

- You misspelled a name.
- You omitted a needed disk drive or path name.
- You entered the parameters without the command name.

Check the spelling on the command line. Make sure that the command, program, or batch file is in the location specified (disk drive and directory path). Then try the command again.

`Bad or missing Command Interpreter`

MS-DOS cannot find the command interpreter, COMMAND.COM. MS-DOS does not start.

If this message appears when you start MS-DOS, COMMAND.COM is not on the boot disk, or a version of COMMAND.COM from a previous version of MS-DOS is on the disk. Place in the floppy disk drive another disk containing the operating system and then reboot the system. After MS-DOS starts, copy COMMAND.COM to the original start-up disk so that you can boot from that disk.

If this message appears while you are running MS-DOS, COMMAND.COM has been erased from the disk and directory you used when starting MS-DOS, or a version of COMMAND.COM from a previous MS-DOS has overwritten the good version. Restart MS-DOS by resetting the system.

If resetting the system does not solve the problem, use a copy of your MS-DOS master disk to restart the computer. Copy COMMAND.COM from this floppy disk to the problem disk.

`Bad or missing filename`

MS-DOS was directed to load a device driver that it could not locate, or an error occurred when DOS loaded the device driver. This message also might mean that a break address for the device driver is out of bounds for the size of RAM in the computer. MS-DOS continues to boot but does not use the device driver.

B

If MS-DOS loads, check your CONFIG.SYS file for the line DEVICE=filename. Make sure that you spell the command correctly and that the device driver is where you specified. If this line is correct, reboot the system.

If the message appears again, copy the file from its original disk to the boot disk and boot MS-DOS again. If the error persists, contact the dealer who sold you the drive because the device driver is bad.

Bad or Missing Keyboard definition file

A DOS warning message, this message means that DOS cannot find KEYBOARD.SYS as specified by the KEYB command. Solving this problem can take several steps. First check to see that KEYBOARD.SYS exists and that it is in the correct path. Then retype the KEYB command. If DOS displays the same message, KEYB.COM or KEYBOARD.SYS may be corrupted.

Bad Partition Table

This message is a FORMAT error message. While using FORMAT, DOS was unable to find a DOS partition on the fixed disk you specified. To correct this problem, run FDISK and create a DOS partition on the fixed-disk drive.

Batch file missing

MS-DOS could not find the batch file it was processing; the batch file was renamed or erased. MS-DOS aborts the processing of the batch file.

If the batch file was renamed, rename it again using its original name. If necessary, edit the batch file to ensure that the file name is not changed again.

If the file was erased, re-create the batch file from its backup file (if possible). Edit the file to ensure that the batch file does not erase itself.

Cannot CHDIR to path - tree past this point not processed

CHKDSK was unable to go to the specified directory. All subdirectories beneath this directory are not verified. To correct this error, run CHKDSK /F.

Cannot CHDIR to root

This message is a CHDIR error message. While checking the tree structure of the directory with CHKDSK, DOS was unable to return to the root directory and did not check the remaining subdirectories. Restart DOS. If DOS continues to display the message, the disk is unusable and must be reformatted.

`Cannot find System Files`

> While using FORMAT /S, you tried to use a drive that does not have the system files in the root directory.

`Cannot load COMMAND.COM, system halted`

> MS-DOS attempted to reload COMMAND.COM, but DOS did not find the command processor, or the area where MS-DOS keeps track of available and used memory was destroyed. The system halts.

> This message can indicate that COMMAND.COM has been erased from the disk and directory you used when starting MS-DOS. Restart MS-DOS. If DOS does not start, your copy of COMMAND.COM has been erased. Restart MS-DOS from the original master disks, and copy COMMAND.COM to your working disk.

> Another possible cause of this message is that a faulty program has corrupted the memory allocation table where MS-DOS tracks available memory. Reboot and then run the same program that was in the computer when the system halted. If the problem occurs again, the program is defective. Contact the dealer or manufacturer who sold you the program.

`Cannot load COMMAND, system halted`

> DOS attempted to reload COMMAND.COM, but DOS did not find the command processor in the directory specified by the COMSPEC=entry, or the area where DOS keeps track of available and used memory was destroyed. The system halts.

> This message can indicate that COMMAND.COM has been erased from the disk and directory you used when starting DOS or that the COMSPEC= entry in the environment has been changed. Restart DOS from your usual start-up disk. If DOS does not start, the copy of COMMAND.COM has been erased. Restart DOS from the DOS start-up or master disk, and copy COMMAND.COM onto your usual start-up disk.

> Another possible cause of this message is that a faulty program has corrupted the memory allocation table where DOS tracks available memory. Run the same program that was in the computer when the system halted. If the problem occurs again, the program is defective. Contact the dealer or manufacturer who sold you the program.

`Cannot perform a cyclic copy`

> When using XCOPY /S, you cannot specify a target that is a subdirectory of the source. Depending on the directory tree structure, you may be able to use a temporary disk or file to overcome this limitation.

B

Cannot read file allocation table

> You were attempting to regain data from good sectors in a bad or defective disk with the RECOVER command when DOS discovered that the file allocation table (FAT) is in a bad sector. Your disk is damaged, and recovering data from the bad sectors may be impossible.

Cannot recover . entry, processing continued

> This message is a CHKDSK warning. While using CHKDSK, this message means that the . entry (the working directory) is defective and cannot be recovered.

Cannot recover .. entry,
Entry has a bad attribute (or link or size)

> This message is an error and warning message. While using CHKDSK, DOS finds that the .. entry (the parent directory) is defective and cannot be recovered.

> If you specified the /F switch, CHKDSK tries to correct the error.

Cannot start COMMAND.COM, exiting

> MS-DOS was directed to load an additional copy of COMMAND.COM but could not. Either the FILES= command in your CONFIG.SYS file is set too low, or you do not have enough free memory for another copy of COMMAND.COM.

> If your system has 256K or more of RAM, and FILES is less than 10, edit the CONFIG.SYS file on your start-up disk to use FILES = 20. Reboot your computer.

> If the problem occurs again, you do not have enough memory in your computer, or you have too many programs competing for memory space. Restart MS-DOS again, and do not load any resident or background programs that you do not need. If necessary, eliminate unneeded device drivers or RAM-disk software. An alternative is to increase the random-access memory in your system.

Configuration too large

> DOS could not load itself because you specified too many FILES or BUFFERS in your CONFIG.SYS file, or you specified too large an environment area (/E switch) to the SHELL command. This problem usually occurs only on systems with less than 256K memory.

B

Restart MS-DOS with a different disk, and edit the CONFIG.SYS file on your boot disk. Lower the number of FILES and BUFFERS or the number after the /E switch in the SHELL command. Restart MS-DOS with the edited disk. An alternative is to increase the memory in your system.

`Content of destination lost before copy`

The source file for COPY was overwritten before the command was completed. A syntax problem caused this error. Restore the source file from your backup disk (or you may be able to use UNDELETE to recover the file).

`Current drive is no longer valid`

You included the current path ($p) in the PROMPT command. MS-DOS attempted to read the current directory for the disk drive and found the drive no longer valid.

If the current disk drive is set for a floppy disk, you do not have a disk in the disk drive. MS-DOS reports a `Drive not ready` error. Press F to fail (the same as A to abort), or I to ignore the error. Then insert a floppy disk into the disk drive or type another drive designation.

An invalid-drive error can also happen if you have a networked disk drive that has been deleted or disconnected. In this case, simply change the current disk to a valid disk drive.

`Data error reading drive d`

DOS could not correctly read the data. Usually the disk has developed a defective spot.

`Disk boot failure`

An error occurred when MS-DOS tried to load itself into memory. The disk contained IO.SYS and MSDOS.SYS, but DOS could not load one of the two files. MS-DOS did not boot.

Start MS-DOS from the disk again. If the error recurs, boot MS-DOS from a disk that you know is good, such as a copy of your MS-DOS start-up or master disk. If this attempt fails, you have a hardware (disk drive) problem. Contact your dealer.

`Disk unsuitable for system disk`

FORMAT/S detected on the floppy disk one or more bad sectors in the area where DOS normally resides. Because DOS must reside on a specific position on the disk and this position is unusable, you cannot use that floppy disk to boot DOS.

B

Reformat the floppy disk. Some floppy disks format successfully the second time. If FORMAT produces this message again, you cannot use that floppy disk as a boot disk.

`Divide overflow`

A program attempted to divide by zero. DOS aborts the program. The program was incorrectly entered, or it has a logic flaw. With a well-written program, this error should never occur. If you wrote the program, correct the error and run the program again. If you purchased the program, report the problem to the dealer or publisher.

`Drive or diskette types not compatible`

When using DISKCOMP or DISKCOPY, you specified drives of different capacities. You cannot DISKCOMP or DISKCOPY from a 1.2M drive to a 360K drive, for example. Retype the command, using like drives.

`Drive not ready`

An error occurred while MS-DOS tried to read or write to the disk drive. For floppy disk drives, the drive door may be open, the disk may not be inserted, or the disk may not be formatted. For hard disk drives, the drive may not be properly prepared, or you may have a hardware problem.

B

`Duplicate filename or File not found`

While using RENAME (or REN), you attempted to rename a file to a name that already existed, or the file you attempted to rename does not exist in the directory. Check the directory for the conflicting names. Make sure that the file name exists and that you have spelled it correctly, and then try again.

`Error in COUNTRY command`

The COUNTRY directive in CONFIG.SYS is improperly phrased or has an incorrect country code or code page number. DOS continues its start-up but uses the default information for the COUNTRY directive.

After DOS has started, check the COUNTRY line in your CONFIG.SYS file. Make sure that the directive is correctly phrased (using commas between country code, code page, and COUNTRY.SYS file) and that its information is correct.

If you detect an error in the line, edit the line, save the file, and restart DOS.

If you do not find an error, restart DOS. If the same message appears, edit your CONFIG.SYS file again, and reenter the COUNTRY directive and delete the old COUNTRY line. The old line may contain some nonsense characters that DOS can (but your text editing program cannot) detect.

`Error in EXE file`

MS-DOS detected an error while attempting to load a program stored in an EXE file. The problem is in the relocation information MS-DOS needs to load the program. This error can occur if the EXE file has been altered in any way.

Restart MS-DOS and load the program again, this time using a backup copy of the program. If the message reappears, the program is flawed. If you are using a purchased program, contact the dealer or publisher.

`Error loading operating system`

A disk error occurred while MS-DOS was loading itself from the hard disk. MS-DOS does not boot.

Restart the computer. If the error occurs after several tries, restart MS-DOS from a floppy disk. If the hard disk does not respond (you cannot run DIR or CHKDSK without an error message), your problem is with the hard disk. Contact your dealer. If the hard disk does respond, use the SYS command to put another copy of MS-DOS on to your hard disk. You also may need to copy COMMAND.COM to the hard disk.

Increase the number of FILES in the CONFIG.SYS file of your start-up disk to 15 or 20. Restart DOS. If the error recurs, you may have a problem with the disk. Use a backup copy of the program and try again. If the backup works, copy it over the problem file.

If an error occurs in the copying process, you have a flawed floppy disk or hard disk. If the problem is the floppy disk, copy the files from the flawed disk to another disk, and reformat or retire the original disk. If the problem is the hard disk, immediately back up your files and run RECOVER on the problem file. If the problem persists, your hard disk may be damaged.

`Error reading directory`

MS-DOS encountered a problem while reading the directory during a format procedure, possibly because bad sectors have developed in the file allocation table (FAT) structure.

B

If the message occurs when DOS reads a floppy disk, the disk is unusable and should be discarded. If the message occurs when DOS reads your hard disk, however, the problem is more serious, and you may have to reformat your disk. Back up your data files on a regular basis to prevent major losses if a problem reading the directory occurs.

Error reading (or writing) partition table

MS-DOS could not read from (or write to) the disk's partition table during FORMAT. This error message indicates that the partition table is corrupted. Run FDISK on the disk and reformat the disk.

EXEC failure

MS-DOS encountered an error while reading a command or program from the disk, or the CONFIG.SYS FILES= command has too low a value.

Increase the number of FILES in the CONFIG.SYS file of your start-up disk to 15 or 20 and then restart MS-DOS. If the error recurs, you may have a problem with the disk. Use a backup copy of the program and try again. If the backup copy works, copy it over the problem copy.

If an error occurs in the copying process, you have a flawed floppy disk or hard disk. If the problem is a floppy disk, copy the files from the flawed disk to another disk, and reformat or retire the original floppy disk. If the problem is the hard disk, immediately back up your files and run RECOVER on the problem file. If the problem persists, your hard disk may have a hardware failure.

File cannot be copied onto itself

You attempted to COPY a file to the same disk, directory, and file name. This message usually indicates that you misspelled or omitted parts of the source or destination drive, path, or file name. This error also can occur when you use wild-card characters for file names. Check your spelling and the source and destination names; then try the command again.

File creation error

MS-DOS or a program could not add a new file to the directory or replace an existing file.

If the file already exists, use the ATTRIB command to check whether the file is marked as read-only. If the read-only flag is set and you want to change or erase the file, use ATTRIB to remove the read-only flag, and then try again.

If the problem is not the read-only flag, run CHKDSK without the /F switch to determine whether the directory is full, the disk is full, or some other problem exists with the disk.

`File not found`

MS-DOS could not find the file you specified. The file is not on the disk or in the directory you specified, or you misspelled the disk drive name, path name, or file name. Check these possibilities, and try the command again.

`Filename device driver cannot be initialized`

In CONFIG.SYS, the parameters in the device driver file name are incorrect, or the DEVICE line is in error. Check for incorrect parameters and for phrasing errors in the DEVICE line. Edit the DEVICE line in the CONFIG.SYS file, save the file, and then restart DOS.

`FIRST diskette bad or incompatible`

`SECOND diskette bad or incompatible`

One of these messages may appear when you use the DISKCOMP command. The messages indicate that either the FIRST (source) or the SECOND (target) floppy disk is unreadable or that the disks you attempted to compare have different format densities.

`Format not supported on drive x:`

The FORMAT command cannot be used on the drive you selected. DOS displays this message if you entered device driver parameters that your computer cannot support. Check CONFIG.SYS for bad DEVICE or DRIVPARM commands.

`General failure reading (or writing) drive d`

This is a catch-all error message when MS-DOS encounters an error it does not recognize. The error usually occurs for one of the following reasons:

- You are using an unformatted disk.
- The disk drive door is open.
- The floppy disk is not seated properly.
- You are using the wrong type of disk in a disk drive, such as formatting a 360K disk in a 1.2M disk drive.

B

Incorrect MS-DOS version

The DOS utility program for the command you just entered is from a different version of MS-DOS.

Find a copy of the program from the correct version of MS-DOS (usually from the MS-DOS master disk) and try the command again. If the floppy disk or hard disk you are using has been updated to hold new versions of the MS-DOS programs, copy those versions over the old ones. If you have more than one version of MS-DOS on your hard disk, make sure that your PATH command refers to the directory with the correct DOS files.

Insert disk with COMMAND.COM in drive d and strike any key when ready

MS-DOS needs to reload COMMAND.COM but cannot find it on the start-up disk.

If you are using floppy disks, the disk in drive A probably has been changed. Place a disk with a good copy of COMMAND.COM in drive A and press any key.

Insert disk with batch file and strike any key when ready

DOS attempted to execute the next command from a batch file, but the disk holding the batch file is not in the disk drive.

Put the disk holding the batch file into the disk drive, and press any key to continue.

Insufficient disk space

The disk does not have enough free space to hold the file being written. All MS-DOS programs terminate when this problem occurs, but some non-DOS programs continue.

If you think that the disk has enough room to hold the file, run CHKDSK to see whether the floppy disk or hard disk has a problem. Sometimes when you terminate programs early by pressing Ctrl + Break, MS-DOS cannot do the necessary clean-up work. When this happens, disk space is temporarily trapped. CHKDSK can "free" these areas.

If you have run out of disk space, free some disk space or use a different floppy disk or hard disk. Then try the command again.

```
Insufficient memory
```

The computer does not have enough free RAM to execute the program or command.

If you have loaded a RAM-resident program like SideKick or DOSKEY, restart MS-DOS, and try the program or command before loading any resident program. If this step fails, remove any unneeded device driver or RAM-disk software from the CONFIG.SYS file, and restart MS-DOS again. If this action fails, your computer does not have enough memory for the program or command. You must increase your random-access memory to run the program or command.

```
Insufficient memory to store macro. Use the DOSKEY command with
the /BUFSIZE switch to increase available memory.
```

Your DOSKEY macros have filled the space set aside for them, and you cannot enter any new macros until you enlarge the memory area (the default is 1,024 bytes). The BUFSIZE switch enables you to increase the amount of memory reserved for DOSKEY macros.

```
Intermediate file error during pipe
```

MS-DOS is unable to create or write to one or both of the intermediate files it uses when piping information between programs. The disk is full, the root directory of the current disk is full, or DOS cannot find the files. The most frequent cause is running out of disk space.

Run the DIR command on the root directory of the current disk drive. Make sure that you have enough free space and enough room in the root directory for two additional files. If you do not have enough room, create room on the disk by deleting or copying and deleting files. You can also copy the necessary files to a different disk.

Another possible cause of this error is that a program is deleting files, including the temporary files DOS uses. If this is the case, you should correct the program, contact the dealer or program publisher, or avoid using the program with piping.

```
Internal stack over flow System halted
```

Your programs and DOS have exhausted the stack (the memory space reserved for temporary use). This problem usually is caused by a rapid succession of hardware devices demanding attention (interrupts). DOS stops, and the system must be turned off and on again to restart DOS.

B

The circumstances that cause this message usually are infrequent and erratic and may not recur. If you want to prevent this error from occurring at all, add the STACKS directive to your CONFIG.SYS file. If the directive is already in your CONFIG.SYS file, increase the number of stacks specified.

`Invalid characters in volume label`

A FORMAT error message and label. You attempted to enter more than 11 alphanumeric characters, or you entered illegal characters (for example, +, =, /, \, or |) when you typed the disk's volume label (the disk name). Retype the volume label, following the proper procedure.

`Invalid COMMAND.COM in drive d`

MS-DOS tried to reload COMMAND.COM from the disk in drive *d* and found that the file was of a different version of MS-DOS. DOS instructs you to insert a disk with the correct version and press a key. Follow these directions.

If you frequently use the disk that was in drive *d*, copy the correct version of COMMAND.COM to that disk.

`Invalid COMMAND.COM, system halted`

MS-DOS could not find COMMAND.COM on the hard disk. MS-DOS halts and must be restarted.

COMMAND.COM may have been erased, or the COMSPEC variable in the environment may have been changed. Restart the computer from the hard disk. If DOS displays a message indicating that COMMAND.COM is missing, the file was erased. Restart MS-DOS from a floppy disk, and recopy COMMAND.COM to the root directory of the hard disk.

If you restart MS-DOS and this message appears later, a program or batch file may be erasing COMMAND.COM. If a batch file is erasing COMMAND.COM, edit the batch file. If a program is erasing COMMAND.COM, contact the dealer or publisher who sold you the program.

`Invalid date`

You entered an impossible date or used the wrong kind of character to separate the month, day, and year. DOS also displays this message if you enter the date with the keypad when the keypad is not in numeric mode.

`Invalid device parameters from device driver`

This message is a FORMAT error message that DOS displays when it finds that the disk partition does not fall on a track boundary. You may have set the DEVICE drivers incorrectly in CONFIG.SYS or attempted to format a hard disk that was formatted with DOS V2.*x* (so the total number of hidden sectors is not evenly divisible by the number of sectors on a track). As a result, the partition does not start on a track boundary.

To correct the error, run FDISK before performing a format, or check CONFIG.SYS for a bad DEVICE or DRIVPARM command.

`Invalid directory`

One of the following errors occurred:

- You specified a directory name that does not exist.
- You misspelled the directory name.
- The directory path is on a different disk.
- You didn't type the path character (\ ) at the beginning of the name.
- You didn't separate the directory names with the path character.

Check your directory names, ensure that the directories do exist, and try the command again.

`Invalid disk change`

The disk in the 720K, 1.2M, or 1.44M disk drive was changed while a program had open files to be written to the disk. DOS displays the message Abort, Retry, Fail. Place the correct disk in the disk drive and press **R** for Retry.

`Invalid drive in search path`

A specification you entered in the PATH command has an invalid disk drive name, or a named disk drive doesn't exist.

Use PATH to check the paths you instructed MS-DOS to search. If you used a nonexistent disk drive name, use the PATH command again to enter the correct search paths. (Or you can just ignore the warning message.)

B

`Invalid drive or file name`

> This error message appears when you have entered the name of a nonexistent disk drive or you mistyped the disk drive, the file name, or both. Check the disk drive name and try the command again.

`Invalid drive specification`

> DOS displays this message when one of the following errors occurs:
>
> - You entered the name of an invalid or nonexistent disk drive as a parameter to a command.
> - You entered the same disk drive for the source and destination, which is not permitted for the command.
> - By omitting a parameter, you defaulted to the same source and destination disk drive.
>
> Check the disk drive names; if the command is missing a parameter and defaulting to the wrong disk drive, explicitly name the correct disk drive.

`Invalid drive specification`
`Specified drive does not exist, or is non-removable`

> One of the following errors occurred:
>
> - You entered the name of a nonexistent disk drive.
> - You named the hard disk drive when using commands for floppy disks only.
> - You omitted a disk drive name and defaulted to the hard disk when using commands for floppy disks only.
> - You named or defaulted to a RAM-disk drive when using commands for a true floppy disk drive.
>
> Certain MS-DOS commands temporarily hide disk drive names while the command is in effect. Check the disk drive name you entered, and try the command again.

`Invalid media or Track 0 bad - disk unusable`

> This message is a FORMAT error message. The disk you are trying to format may be damaged, but often a disk does not format the first time. Try to format the disk again; if the same message appears, the disk is bad and should be discarded.

`Invalid number of parameters`

> You have entered too few or too many parameters for a command. One of the following errors occurred:

B

- You omitted required information.
- You excluded a colon immediately after the disk drive name.
- You put a space in the wrong place or omitted a needed space.
- You omitted a slash (/) in front of a switch.

`Invalid parameter  Incorrect parameter`

At least one parameter you entered for the command is not valid. One of the following errors occurred:

- You omitted required information.
- You omitted a colon immediately after the disk drive name.
- You put a space in the wrong place or omitted a needed space.
- You didn't add a slash (/) in front of a switch.
- You used a switch the command does not recognize.

`Invalid parameter combination`

When you entered an MS-DOS command, you typed parameters that conflict. Retype the command, using only one of the conflicting switches.

`Invalid partition table`

When you started DOS from the hard disk, DOS detected a problem in the hard disk's partition information.

Restart MS-DOS from a floppy disk. Back up all files from the hard disk (if possible), and run FDISK to correct the problem. If you change the partition information, reformat the hard disk, and restore all its files.

`Invalid path`

One of the following errors has occurred:

- The path name contains illegal characters.
- The path name has more than 63 characters.
- One of the directory names within the path is misspelled or does not exist.

Check the spelling of the path name. If needed, find a directory listing of the disk to ensure that the directory you specified does exist and that you have the correct path name. Be sure that the path name contains no more than 63 characters. If necessary, change the current directory to a directory "closer" to the file to shorten the path name.

B

Invalid path or file name

> You entered a directory name or file name that does not exist, used the wrong directory name (a directory not on the path), or mistyped a name. COPY aborts when it encounters an invalid path or file name. If you used wild-card characters in a file name, COPY transfers all valid files before it issues the error message.

> Check to see which files were transferred. Determine whether the directory and file names are spelled correctly and whether the path is correct. Then try the command again.

Invalid time

> You entered an impossible time or used the wrong kind of character to separate the hours, minutes, and seconds. DOS also displays this message if you enter the time with the keypad when the keypad is not in numeric mode.

Invalid Volume ID

> This FORMAT error message occurs when you enter an incorrect volume label (the name of the disk drive) during the formatting of a fixed (hard) disk. DOS aborts the format attempt.

> To view the volume label of the disk, type **VOL** at the prompt and press Enter. Then try the command again.

Lock violation

> With the file-sharing program (SHARE.EXE) or network software loaded, one of your programs attempted to access a locked file. First try **Retry**; if unsuccessful, try **Abort** or **Fail**. (If you choose abort or fail, however, you lose any data in memory.)

Memory allocation error
Cannot load COMMAND, system halted

> A program damaged the area where MS-DOS keeps track of available and used memory. You must restart MS-DOS.

> If this error occurs again with the same program, use a backup copy of the program. If the problem persists, the program has a flaw. Contact the dealer or program publisher.

MIRROR cannot operate with a network

> MIRROR cannot save file reconstruction information because your computer's hard disk is currently redirected to a network.

B

```
Missing operating system
```

The MS-DOS hard disk partition does not have a copy of MS-DOS on it. MS-DOS does not boot.

Start MS-DOS from a floppy disk. Use the SYS C: command to place DOS and COMMAND.COM on the hard disk. If this command fails to solve the problem, back up the existing files (if any) from the hard disk, and then issue the FORMAT /S command to put a copy of the operating system on the hard disk. If necessary, restore the files that you backed up.

```
No free file handles
Cannot start COMMAND, exiting
```

MS-DOS could not load an additional copy of COMMAND.COM because no file handles were available.

Edit the CONFIG.SYS file on your start-up disk to increase the number of file handles (using the FILES command) by five. Restart DOS and try the command again.

```
Non-DOS disk
```

MS-DOS does not recognize the disk format as a DOS disk. This disk is unusable. Abort and run CHKDSK to learn whether any corrective action is possible. If CHKDSK fails, an alternative is to reformat the disk.

Reformatting destroys any information remaining on the disk. If you have disks from another operating system, this disk was probably formatted under the other operating system and should not be reformatted.

```
No room for system on destination disk
```

The floppy disk or hard disk was not formatted with the necessary reserved space for MS-DOS. You cannot put the system on this floppy disk without first copying all the disk's data to another disk and then reformatting the disk.

```
Non-System disk or disk error Replace and strike any key when
ready
```

Your floppy disk or hard disk does not contain MS-DOS, or a read error occurred when you started the system. MS-DOS does not boot. If you are using a floppy disk system, put a bootable disk into drive A and press any key.

The most frequent cause of this message on hard disk systems is that you left a nonbootable floppy disk in drive A with the drive door closed.

B

Open the door to drive A and press any key. MS-DOS boots from the hard disk.

`No paper`

The printer is out of paper or is turned off.

`No system on default drive`

SYS could not find the system files. Insert a disk containing the system files, such as the DOS disk, and enter the command again.

`No target drive specified`

You did not specify a target drive when you typed a BACKUP command. Retype the command, using first a source and then a target disk drive.

`Not enough memory`

The computer does not have enough free memory to execute the program or command.

If you loaded a RAM-resident program, such as SideKick or DOSKEY, restart MS-DOS, and try the program or command again before loading any resident program. If this method fails, remove any unneeded device driver or RAM-disk software from the CONFIG.SYS file, and restart MS-DOS.

If this procedure fails, your computer does not have enough memory for this operation. You must increase your RAM to run the program or command.

B

`Not ready`

A device is not ready and cannot receive or transmit data. Check the connections, check that the power is on, and check whether the device is ready. For floppy disk drives, check that the disk is formatted and properly seated in the disk drive.

`Out of environment space`

DOS is unable to add any more strings to the environment from the SET command. The environment cannot be expanded. This error occurs when you load a resident program, such as DOSSHELL, MODE, PRINT, GRAPHICS, or SideKick.

`Out of memory`

The amount of memory is insufficient to perform the operation you requested. This error occurs in EDIT, the DOS full-screen text editor.

Packed File Corrupt

> A program file did not successfully load into the first 64K of memory. This error can occur when a packed executable file is loaded into memory. Use the LOADFIX command to load the program above the first 64K.

Parameters not supported

Parameters not supported on drive

> You entered parameters for a command that do not exist, are not supported by MS-DOS, or are incompatible with the disk drive that you selected.

Path not found

> A file or directory path you named does not exist. You misspelled the file name or directory name, or you omitted a path character (\ ) between directory names or between the final directory name and file name. Another possibility is that the file or directory does not exist where you specified. Check these possibilities and try again.

Path too long

> You entered a path name that exceeds the 63-character limit of MS-DOS. The name is too long, or you omitted a space between file names. Check the command line. If the phrasing is correct, change to a directory that is closer to the file you want, and try the command again.

B

Program too big to fit in memory

> The computer does not have enough memory to load the program or command you invoked. Type **exit** to ensure that you do not have another application program in memory.

> If you have any resident programs loaded (such as DOSKEY), restart MS-DOS, and try the command again without loading the resident programs. If this message appears again, reduce the number of buffers (BUFFERS=) in the CONFIG.SYS file, and eliminate unneeded device drivers or RAM-disk software. Restart MS-DOS. If these actions do not solve the problem, your computer lacks the memory needed to run the program or command. You must increase the amount of RAM in your computer to run the program or command.

`Read fault error reading drive d`

MS-DOS was unable to read the data, usually from a hard disk or floppy disk. Check that the disk drive door is closed and that the disk is properly inserted.

`Same parameter entered twice`

You duplicated a switch when you typed a command. Retype the command, using the parameter only once.

`Sector not found error reading drive d`

The disk drive was unable to locate the sector on the floppy disk or hard disk platter. This error is usually the result of a defective spot on the disk or defective drive electronics. Some copy-protection schemes use a defective spot to prevent unauthorized duplication of the disk.

`Seek error reading (or writing) drive d`

The disk drive could not locate the proper track on the floppy disk or hard disk. This error is usually the result of a defective spot on the floppy disk or hard disk platter, an unformatted disk, or drive electronics problems.

`SOURCE diskette bad or incompatible`

The disk you are copying is damaged or is the wrong format (for example, a high-density 5 1/4-inch disk in a double-density 5 1/4-inch disk drive). DOS cannot read the disk.

`Syntax error`

You phrased a command improperly for one of the following reasons:

- You omitted needed information.
- You entered extra information.
- You put an extra space in a file name or path name.
- You used an incorrect switch.

Check the command line for these possibilities, and try the command again.

```
TARGET diskette bad or incompatible
```

```
Target diskette may be unusable
```

```
Target diskette unusable
```

> This DISKCOPY message indicates that a problem exists with the target disk, MS-DOS does not recognize the format of the target disk in the drive, or the disk is bad.

> Check that the disk is the same density as the source disk, run CHKDSK on the target disk to determine the problem, or try to reformat the disk before proceeding with the disk copy operation.

```
TARGET media has lower capacity than SOURCE
Continue anyway (Y/N)?
```

> This DISKCOPY warning message informs you that the target disk can hold fewer bytes of data than the source disk. The most likely cause is a target disk with bad sectors. If you type a **Y** (Yes), some of the data on the source disk may not fit on to the target disk.

> To avoid the possibility of an incomplete transfer of data, type **N** and use a disk that has the same capacity as the source disk, or you can use the COPY *.* command to transfer the files if you are not copying hidden files.

```
There is not enough room to create a restore file
You will not be able to use the unformat utility Proceed with
Format (Y/N)?
```

> FORMAT has determined that the disk lacks sufficient room to create a RESTORE file. Without this file, you cannot use the UNFORMAT command to reverse the format that you are attempting.

```
Unable to create directory
```

> You or a program attempted to create a directory, and one of the following errors occurred:

> - A directory of the same name exists.
> - A file of the same name exists.
> - You tried to add a directory to a root directory that is full.
> - The directory name has illegal characters or is a device name.

B

List the directories of the disk. Make sure that no file or directory with the same name already exists. If you are adding the directory to the root directory, remove or move (copy and then erase) any unneeded files or commands. Check the spelling of the directory, and ensure that the command is properly phrased.

`Unable to load MS-DOS Shell, Retry (y/n)?`

DOS could not load the Shell. You may have another program in memory, and the Shell will not fit into memory, or the DOS Shell program itself may be corrupted.

Exit the program and try to load the Shell. If the Shell still doesn't load, it is probably corrupt. Reboot your system and load the Shell. If the same error message appears, copy the Shell from a backup disk to your hard disk.

`Unable to write BOOT`

FORMAT could not write to the first (or BOOT) track or DOS partition of the disk that you are formatting because one of these areas is bad. Discard the bad disk, insert another unformatted disk, and try the FORMAT command again.

`Unrecognized command in CONFIG.SYS`

MS-DOS detected an improperly phrased directive in CONFIG.SYS. The directive is ignored, and MS-DOS starts. Examine the CONFIG.SYS file, looking for improperly phrased or incorrect directives. Edit the line, save the file, and restart MS-DOS.

`Unrecoverable read error on drive x side n, track n`

MS-DOS could not read the data at the described location on the disk. (MS-DOS makes four attempts before generating this message.) Copy all files on the questionable disk to another disk, and try the command again, first with a new disk and then with the backup disk. If the original disk cannot be reformatted, discard it.

`Unrecoverable write error on drive x Side n, track n`

MS-DOS was unable to write to a disk at the location specified. Try the command again. If the same error occurs, the target disk is damaged at that location.

If the damaged disk contains important data, copy the files to an empty formatted disk, and try to reformat the damaged disk. If the disk is bad, discard it.

B

Write fault error writing drive d

> MS-DOS could not write the data to this device. You may have inserted the floppy disk improperly or left the disk drive door open. Another possibility is an electronics failure in the floppy or hard disk drive. The most frequent cause is a bad spot on the disk.

Write protect error writing drive d

> You tried to write data to a disk that is write-protected. Use a different disk, or remove the write-protection tab on a 5-1/4 inch disk, or slide the write-protection tab to the write-enable position on a 3-1/2 inch disk.

B

Index

Symbols

, (comma), 68
* (asterisk) wild card, 73
... (ellipsis), 48, 99
/ (slash), 59, 69
: (colon), 40, 68, 99
; (semicolon), 175
< (redirect input), 164
> (greater than) symbol, 40
> redirect output symbol, 164
>> (redirect output and add text), 164
? (question mark) wild card, 73
\ (backslash), 68, 94, 99
| (pipe), 165

A

absolute references, 115
adapters, 11-12, 16
address bus, 22
applications programs, 30, 34, 37
    directories, 105
    interface, 37
ASCII (American Standard Code for Information Interchange)
    files, 148
    format, 181
asterisk (*) wild card, 73
AUTOEXEC.BAT file, 36, 174, 184-189

B

backslash (\), 68, 94, 99
BACKUP command, 132-133, 136-138
backups, 136
    AUTOEXEC.BAT, 188-189
    CHKDSK command, 152-153
    disks, 121, 153
    DOS disks (installation), 199-201
    files, 138-140
    floppy disks, 121-122, 134-135
    frequency, 133-134
    full backup, 133, 137
    hard disks, 133-134
    restoring, 131-133
    selected directories, 138
bad sectors (disks), 82, 148, 160
basic input/output system, see BIOS
BAT (batch file) extension, 179
\BATCH directory, 103
batch files, 30, 103, 173-174
    creating, 181-184
    directories, 181
    guidelines, 183
    names, 180
    path, 175
    running, 183-184
    saving, 181
    statements, 179
    viewing contents, 180
    see also text files

BBSs (bulletin board systems), 22, 123
BIOS (basic input/output system), 30, 33
bits, 7, 22
boilerplates, 119
booting, 30, 38-40, 198
buffers, 69, 174, 190-191
BUFFERS command (CONFIG.SYS file), 190
bugs, 123
bulletin board systems, *see* BBSs
buses, 22
bytes, 7, 22

C

C:\> (DOS prompt), 33
canceling menus, 47-48
cathode ray tube, *see* CRT
CD command, 106
central processing unit, *see* CPU
CGA (color graphics adapter), 11
characters (metastrings), 177-178
CHDIR command, 106
check boxes (dialog boxes), 48
child directories, 97
chips, 8
CHKDSK command, 134, 149-154
clearing
   display, 154-155
   FAT, 126
   screen, 154-155
   see also deleting
clicking mouse, 31
CLS command, 154-155
clusters (disks), 148, 150
cold boot, 30, 38-39
colon (:), 40, 68, 99
color (display adapters), 12
color graphics adapter (CGA), 11
COM (command) file extension, 34
command elements, 62-69
command files, 34
command interpreter, 32-33
command line, 58, 69-70

command processor, 59
COMMAND.COM file, 32, 36
   DOS prompt, displaying, 33
   root directory, 103
commands, 30, 58-68
   / (slash mark), 59
   as programs, 60
   AUTOEXEC.BAT file, 185-187
   BACKUP, 132-133, 136
   CD, 106
   CHDIR, 106
   CHKDSK, 134, 149-154
   CLS, 154-155
   COPY, 85-86
   COPY CON, *see* COPY CON
   correcting typing mistakes, 69-70
   custom, 165-167
   DEL, 68
   DELETE, 87
   delimiters, 58, 59
   dialog boxes, 48-49
   DIR, 70, 116-117
   DISKCOMP, 59
   DISKCOPY, 59, 120-122
   ellipsis (...), 48
   ERASE, 68, 126
   executing, 68-70
   external, 33, 60
   filters, 165-167
   FORMAT, 75, 82, 126
   HIMEM, 207
   internal, 33, 60
   LABEL, 158-159
   line-editing, 69
   MD, 106
   MEM, 149
   MIRROR, 127-128
   MKDIR, 106
   Options menu (Shell), 50
   parameters, 58-60, 68-69, 94
   PATH, 101, 174-177
   pipes, 165-167
   PROMPT, 177-179
   RD, 108
   RECOVER, 159-161
   RENAME, 87-88

RESTORE, 84, 132-133, 139-140
RMDIR, 108
selecting from menus, 41
switches, 58, 61-68
syntax, 58-61
TREE, 102
TYPE, 88
typing name, 68
UNDELETE, 128
UNFORMAT, 130-131
VER, 155-156
VERIFY, 156-157
VOL, 157-158
communications, 21
CONFIG.SYS file, 174, 189
    BUFFERS command, 190
    creating, 191-192
    editing, 191-192
    FILES command, 190
configuration files, 189
    see also CONFIG.SYS
console, 148
    see also COPY CON
control bus, 22
COPY command, 85-86
COPY CON, 162-163, 182-183
copying
    disks, 59, 120-122
    files, 85-86, 117-119
        backups, 139-140
        between directories, 117-118
        in same directory, 118
    from console, *see* COPY CON
CPU (central processing unit), 7
creating
    AUTOEXEC.BAT file, 188-189
    batch files, 181-184
    CONFIG.SYS file, 191-192
    directories, 97, 106
    path name, 99
    root directory, 97
    text file, 163
CRT (cathode ray tube), 10
current
    directory, 106-107, 114
    drive, 71

cursor, 30
    mouse, 20
    selection cursor, 31
    selection cursor (Shell), 44

D

\DATA directory, 104
data bus, 22
data loss, 122-127
date/time stamps (files), 73
defaults
    AUTOEXEC.BAT file, 185
    startup directory, 103
defragmenting disks, 148
DEL command, 68
DELETE command, 87
delete-tracking file, 127
deleting
    directories, 107-108
    files, 119-120
        floppy disks, 87
        recovering, 126-142
        undeleting, 128-129
    volume labels (disks), 158
    see also clearing
delimiters, 58-59, 68
density (floppy disks), 77
destination disks, 114, 120
destination files, 85, 114
devices, 148
    console, 148
    names, 161, 163
    output, redirection, 161-165
dialog boxes, 31, 41, 48-50
DIR command, 70, 74, 116-117
directories, 70, 94-95
    applications software, 105
    backups, selected directories, 138
    \BATCH, 103
    batch files, 181
    child, 97
    creating, 106
    current, 106-107, 114
    \DATA, 104
    default at startup, 103

deleting, 107-108
displaying with TREE command, 102
\DOS, 103
files
    deleting, 119-120
    FIND filter, 166
    listing, 71
    viewing, 116-117
floppy disks, 106
hard disks, 105-108
hierarchical structure, 94-95
\KEEP, 105
\MISC, 105
navigation, 95-97
parent, 97
PATH command, 175
restoring selected directories, 141
root directory, 100-103, 107
subdirectories, 94-97, 102-105
\TEMP, 104
tree structure, 94-96
\UTIL, 103
\UTILITY, 103
directory specifier, 94, 98-99
Directory Tree area (Shell), 44, 51
disk drives
    floppy disks, 78-79
    heads, 125
disk operating system, *see* DOS
disk space, 149-154
Disk Utilities program group (Restore Fixed Disk option), 140
DISKCOMP command, 59
DISKCOPY command, 59, 120-122
disks, 7, 17-19, 36-37
    allocation unit, 58
    backups, 121, 131-133
    clusters, 150
    copying, 59, 120-122
    defragmenting, 154
    destination disks, 114, 120
    drives, *see* drives; hard disks
    files, 18
    floppy, *see* floppy disks
    formatting, 58, 97, 130-131

fragmentation, 148, 153-154
hard, *see* hard disks
installation, backups, 199-201
microfloppies, 18
minifloppies, 18
overwriting, 114
reading, 17
sectors, 58, 82, 148, 160
serial number, 157
source disks, 114, 120
tracks, 17, 58
unformatting, 58, 130-131
verbose display mode, 150
viewing with CHKDSK command, 150
volume label, 58, 80, 157-159
working disks, 198
writing to, 17
disk drives, *see* drives; hard disks
display, 7
    clearing, 154-155
    display adapters, 11-12
    pixels, 10
    see also monitors
displaying
    directories with TREE command, 102
    DOS prompt (COMMAND.COM), 33
    files, sorting by extension, 50
    system files, hidden, 50
displays, 10-12
DOC (text) file extension, 34
\DOS directory, 103
DOS, 9
    versions, 155-156
        conflicts, 142
        upgrading, 201, 204-206
DOS Editor, 161
DOS prompt
    C:\>, 33
    command line, 58
    commands, 68-70
    customizing, 178-179
    displaying in COMMAND.COM, 33
    editing, 177-179

DOS Shell, *see* Shell
DOSSHELL.INI file, 36, 209
dot-matrix printers, 20
double-sided floppy disks, 18, 77
dragging mouse, 31
drive letters (Shell), 44
drive specifier, 175
drives, 18
    current, changing, 71
    floppy disks, 40
    see also hard disks
duplicating, *see* copying

E

EDIT text editor, 181
editing
    CONFIG.SYS file, 191-192
    DOS prompt, 177-179
    volume label (disks), 158-159
editing keys in command line, 69-70
EGA (enhanced graphics adapter), 11
EGA.CPI, 36
ellipsis (...), 48, 99
encoding disks, 76
entries (dialog boxes), 50
ERASE command, 68, 126
erasing, *see* clearing
error messages, 211-235
EXE (executable) file extension, 34
executable files, 34
executing commands, 69
executing programs, 34
exiting Shell, 47
expansion slots, 16
extended graphics array adapter
  (XGA), 11
extended keyboard, 12-15
extensions (file names), 34, 72-73
    BAT (batch file), 179
    COM (command), 34
    DOC (text), 34
    examples, 35
    EXE (executable), 34
    sorting display, 50
    TXT (text), 34
external commands, 33, 60

F

FAT (file allocation table), 75,
  126-128
File Display Options command
  (Options menu), 50
file management, 36-37
file names (extensions), 34, 72-73
    BAT (batch file), 179
    COM (command), 34
    DOC (text), 34
    examples, 35
    EXE (executable), 34
    TXT (text), 34
files, 7, 18
    ASCII, 148
    AUTOEXEC.BAT, *see*
      AUTOEXEC.BAT
    backups, 138-140
    batch files, *see* batch files
    command files, 34
    COMMAND.COM, 36
    CONFIG.SYS, *see* CONFIG.SYS file
    configuration, 189
        see also CONFIG.SYS file
    copying, 85-86, 117-119
    date/time stamps, 73
    deleting, 119-120
        tracking, 127
        recovering, 126-142
        undeleting, 128-129
    destination files, 85, 114
    directories
        FIND filter, 166
        listing, 71
        viewing, 116-117
    DOSSHELL.INI, 36, 209
    EGA.CPI, 36
    executable, 34
    FIND.EXE, 36
    floppy disks, deleting, 87
    fragmentation, 148, 153-154
    hidden, 33
    input/output system, 33
    KEYBOARD.SYS, 36

listing, 50
   multiple, 75
   one screen at a time, 72
   with FIND filter, 166
names, 36
noncontiguous, 148
overwriting, 114
programs, 34
README.DOC, 34
README.TXT, 207-209
recovering, 126-142, 159-161
renaming, 87-88
restoring, 140-141, 160
scrolling, 167
searches, starting, 100-101
SELECT.HLP, 36
size, 73
sorting by extension, 50
source files, 85, 114
temporary, 104
text, *see* text files
undeleting, 128-129
unreadable, 160
utility files, 33-34
verifying, 156-157
wild cards, 73-75
Files area (Shell), 44
FILES command (CONFIG.SYS), 190
filters, 148, 165-167
FIND filter, 166
FIND.EXE, 36
fixed disks, *see* hard disks
floppy disks, 17-19, 75-78
   backups, 121-122, 134, 137
   density, 77
   directories, creating, 106
   double-sided, 18, 77
   drives, 40, 78-79
   encoding, 76
   files, deleting, 87
   formatting, 79-83
   installing DOS 5, 203-204
   labels, 115
   microfloppies, 18
   minifloppies, 18
   precautions, 19

   sectors, 76, 82
   single-sided, 77
   tracks, 76
   unformatting, 80
   unusable areas, 83
   upgrading, 207
   volume labels, 80
   write-protection, 79
FORMAT command, 75, 82-84, 126
formatting
   disks, 58, 97, 130-131
   files, text files, 181
   floppy disks, 79-83
   hard disks, 84, 198, 201
fragmentation, 148, 153-154
full backup, 133, 137
full restore, 140
function keys, 13, 15

G–H

greater than (>) symbol, 40
GUI (graphical user interface), 30

hard disks, 17-19, 95
   backups, 133-134
   buffers, 191
   clusters, 148
   directories, 105-108
   files, copying backups, 139-140
   formatting, 84, 198, 201
   head-parking program, 114
   installing DOS 5, 201-203
   partitions, 198
   self-parking heads, 114
   upgrading, 205-206
hardware, 8-10
   display, 7
   expansion slots, 16
   failures, preventing, 124-126
   modems, 7
   monitor, 7
   motherboard, 16
   peripherals, 7, 19-22
   printers, 20-21
   screen, 7
   system unit, 16-17

head-parking program, 114
heads (disk drives), 125
Help, 53
Hercules monochrome graphics
adapter (MGA), 11
hidden files, 33, 50
hierarchical directories, 94-95
HIMEM command, 207
history of computers, 8

I

IBM PCs, 8
icons, 31
inkjet printer, 21
input, 7
input devices, 12
input/output system, 33
installation
disks, 198-201
README.TXT file, 207-209
reinstalling, 209-210
requirements, 198
SETUP procedure, 201-203
to floppy disks, 203-204
to hard disk, 201-203
uninstalling, 198, 209
upgrading, 201, 204-206
integrated circuits, 8
interfaces, 30, 37
internal commands, 33, 60

K

K (kilobytes), 7
\KEEP directory, 105
key combinations, 15
keyboard, 12-15
editing keys (command line),
69-70
selecting menu options, 47
Shell, 46
special keys, 13-15
KEYBOARD.SYS, 36

L

LABEL command, 158-159
labels (floppy disks), 115
laser printers, 20
/LIST switch (UNDELETE
command), 128
list boxes (dialog boxes), 49
listing files
directories, 71
FIND filter, 166
multiple, 75
one screen at a time, 72
loading Shell, 43
lost clusters (disks), 150
low-level format, 198, 201

M

M (megabytes), 7
Mace Utilities, 126
MD command, 106
MDA (monochrome display
adapter), 11
megabytes (M), 7
MEM command, 149
memory
availability, 149
requirements (installation), 198
memory report utility, 149
menu bar (Shell), 43
menus
canceling, 47-48
commands, selecting, 41, 46
Help, 53
options, selecting, 47
pull-down, 31, 47
metastrings, 174, 177-178
MGA (Hercules monochrome
graphics adapter), 11
microcomputer, 8
microfloppies, 18
microprocessor, 8
minifloppies, 18
MIRROR command, 127-128
\MISC directory, 105

MKDIR command, 106
modems, 7, 21-22
monitors, pixels, 10
  see also display
monochrome display adapter
  (MDA), 11
MORE filter, 167
motherboard, 11, 16
mouse, 19-20
  clicking, 31
  dragging, 31
  pointer, 20, 31, 44
  selecting menu items, 46-47
  Shell, 41, 45
MS-DOS, *see* DOS

N

names
  batch files, 180
  devices, 161-163
  files, 34-36
navigating, 95-97
Norton Utilities, 126

O

on-line help, 53
operating system, 9, 31-34
option buttons (dialog boxes), 49
options, 41, 47
Options menu, 50
output, 7
  filters, 148
  piping, 148
  redirection, 148, 161-165
overwriting, 114

P

parameters, 58, 60
  adding to commands, 68-69
  delimiters, 68
  directory specifier, 94
parent directories, 97
partitions (hard disk), 198

path, 58, 98-102, 115-116, 174-177
  absolute reference, 115
  batch files, 175
  changing, 176-177
  drive specifier, 175
  path name, 94, 98-99
  relative references, 115
  search path, 175
  viewing, 101-102
PATH command, 101, 174-177
PC DOS, *see* DOS
PC Tools Deluxe, 126
peripherals, 7, 19-22
  see also hardware
picture element (pixel), 10
pipes, 148, 165-167
pixels, 10
pointers, 31
  mouse, 19-20
  Shell, 44
ports (printers), 20
printers, 20-21
Program area (Shell), 44
programs, 9, 23, 30, 34
  commands as, *see* external
    commands
  executing, 34
  see also software
PROMPT command, 177-179
prompt view (Shell), 40
prompts, *see* DOS prompt
pull-down menus, 31, 47

Q–R

question mark (?) wildcard, 73

RAM (random-access memory), 23,
  149
RD command, 108
read-only memory, *see* ROM
reading disks, 17
README.DOC file, 34
README.TXT file, 207-209
RECOVER command, 159-161

recovering
  disks, 130-131
  files, 126-142, 159-161
redirection (output), 148, 161-165
reinstalling, 209, 210
relative references, 115
removing, *see* deleting
RENAME command, 87-88
renaming files, 87-88
requirements (installation), 198
Reset button, 39
resolution, 10, 12
RESTORE command, 84, 132-133, 139-140
Restore Fixed Disk option (Disk Utilities program group), 140
restoring
  backups, 131-133
  directories, 141
  files, 140-141, 160
RMDIR command, 108
ROM (read-only memory), 23
  boot instruction set, 39
  hidden files, 33
root directory, 102-103
  \ (backslash), 97
  batch files, 181
  changing to, 107
  COMMAND.COM file, 103
  creating, 97
  viewing, 100
running batch files, 183-184

S

safe formatting, 80, 130
safeguards (data loss), 122-126
saving
  batch files, 181
  temporary files, 104
screen clearing, 154-155
  see also display; monitor
scroll bars (Shell), 44, 50-52
scrolling, 51
  control, 72
  files, 167
search path, 175

searches, 73-75, 100-101
sectors (disks), 58, 76
  bad sectors, 82, 148, 160
SELECT.HLP, 36
selecting
  commands, 41
  menu items, 46-47
  Shell items, 45
selected backup, 137
selection cursor, 31, 44
selective restore, 141
self-parking heads (hard disks), 114
semicolon (;) drive specifier, 175
serial number (disks), 157
SETUP procedure (installation), 201-203
Shell, 30, 40-53
  backups, 137
  exiting, 47
  loading, 43
  Program area, 44
  prompt view, 40
  RESTORE command, 140
  scroll bars, 44, 50-52
  settings, preserving, 209-210
  Shell view, 41-42
single-sided floppy disks, 77
slash (/), 59, 69
software, 8-10
  failures, preventing, 123-124
  utilities, 126
  viruses, 123
  see also programs
SORT filter, 166-167
sorting files by extension, 50
source disks, 114, 120
source files, 85, 114
special characters (metastrings), 177-178
special keys, 13-15
starting searches, 100-101
startup defaults, 103
  see also booting
statements in batch files, 179
static electricity, 114, 125
status line (Shell), 44
status reports (disk space), 150

subdirectories, 94-97, 102-105
surge suppressors, 114, 124
switches, 58, 61-68
    BACKUP command, 136, 138
    CHKDSK command, 150, 152
    RESTORE command, 139
    TREE command, 102
    UNDELETE command, 128
syntax, 58-61
system files, 50
system unit, 16-17

T

target, *see* destination
\TEMP directory, 104
temporary files, 104
text boxes (dialog boxes), 48
text editors, *see* EDIT text editor
text files, 34
    ASCII format, 181
    creating with COPY CON, 163
    README.DOC, 34
    redirection to printer, 165
    viewing contents, 88
    see also batch files
title bar (Shell), 43
tracking FAT, 128
tracks (disks), 17, 58, 76
TREE command, 102
tree structure (directories), 94-96
TXT (text) file extension, 34
TYPE command, 88
typeovers in dialog box entries, 50
typing command name, 68

U

UNDELETE command, 128
UNFORMAT command, 130-131
unreadable files, 160
upgrading, 201, 204-207
\UTIL directory, 103
utilities, 33-34
    backups, 132
    directory, 103
    DISKCOPY, 120

Mace Utilities, 126
MEM, 149
MIRROR, 128
Norton Utilities, 126
PC Tools Deluxe, 126
\UTILITY directory, 103

V

VER command, 155-156
verbose display mode, 150
VERIFY command, 156-157
VGA (video graphics array adapter), 11
video display, *see* display; monitors
viewing
    AUTOEXEC.BAT file, 188
    batch files contents, 180
    disk contents with CHKDSK command, 150
    files in other directories, 116-117
    path, 101-102
    root directory, 100
    text files, 88
viruses, 123
VOL command, 157-158
volume label (disks), 58, 80, 157-159

W

warm boot, 30, 38-40
wild cards, 58, 73-75
    * (asterisk), 73
    ? (question mark), 73
    COPY command, 85
    DIR command, 74
    UNDELETE command, 128
working disks, 198
write-protecting floppy disks, 79, 83
writing to disks, 17

X–Y–Z

XGA (extended graphics array adapter), 11